THE INS AND OUTS OF UPS AND DOWNS

THE INS AND OUTS OF UPS AND DOWNS

By

Ira L. Fox, M.D.

LIBRA PUBLISHERS, INC.

This book is dedicated to my daughters,
Beth and Leslie;
to all of my patients who have
taught me so much; and to my friends
who have stood by me during the process

First Edition

Library of Congress Catalog No. 82-090730

Libra Publishers, Inc.
391 Willets Road
Roslyn Heights, N.Y. 11577

Manufactured in the United States of America

ISBN 0-87212-170-4

CONTENTS

FOREWORD

This book aims to present, as clearly and concisely as possible, the major common factors which cause us all to undergo mood changes and mood swings which seem beyond our conscious control. My impression is that the inability to control moods is the most common presenting complaint of psychiatric patients. It may, in fact, be one of the most common problems among the American population in general. It is surely among the most common complaints among couples whose relationships are in distress. The intensity and frequency of mood changes are certainly exemplified by the teenage and young adult years.

The Ins and Outs of Ups and Downs explains how the precepts of psychology, general medicine, and psychopharmacology can dovetail to help you become healthier, more comfortable, and more consistently stable emotionally.

The Ins and Outs of Ups and Downs (1) shows you how to achieve and maintain a consistent mood, (2) illustrates how you can improve your relationships with others when your moods become stable, (3) shows you how to be less moody and less easily hurt by other people, and (4) demonstrates clearly that you can remain sensitive to other people and their needs without being overly sensitized by their moods and attitudes.

This book is not presented as an experimental research document. I am not involved in research. I am a clinician—a practitioner of inpatient and outpatient general psychiatry.

This book reports my clinical observation and evaluations of a large and diverse group of patients. A wide range of ethnic, religious, socioeconomic, educational, and professional backgrounds have been represented in my practice. Thus, my conclusions in these pages transcend maleness and femaleness. They transcend individual backgrounds and ages.

Many of the statements in this book are controversial. My sincerest hope is that this text will stimulate a great deal of activity, not only within the psychiatric community, but among the general population as well. It is time that the public was made aware of the simple treatments which are available and which may, in fact, be of such general applicability as to be able to help the vast majority of individuals who are moody, explosive, impulsive, and overwhelmingly sensitive to others.

Please read *The Ins and Outs of Ups and Downs* critically, but with an open mind. As with any presentation of provocative information, some people will agree with my premise and some will disagree. I hope that those who disagree with me will be vocal enough in addressing themselves to the issues, so that more and more viewpoints can be brought forth for public inspection and evaluation.

I am first a psychiatrist, and only second an author. I have had a great deal of help in making my ideas readable and understandable. When this book was in its mere infancy, Mrs. Jeanne Kessler offered many suggestions and early revisions. Ms. Patricia Seay assisted me with the overwhelming majority of the revisions and technical work. Pat spent untold hours laboring over revision after revision. Without her help this book would never have reached completion. Her interest and steadfast support kept me going at times when giving up would have been much easier. I would also like to thank my office staff, especially my office manager, Mrs. Donna Symons, for their support in typing the manuscript for this book.

INTRODUCTION

Were you ever told that "anybody can do it," but you couldn't? You may have felt that you should have been able to, but—you just couldn't make yourself do it.

Many "self-help" and "assertiveness training" books have come on the market in recent years. These books have been written to help the average person deal with the stresses and strains of daily living. These manuals of coping techniques are supposedly of such a nature that anybody with reasonable intelligence can utilize them. But everyone can't! Many people, even some of above-average intelligence, are unable to use these techniques because their moods are inconsistent. The most efficient of techniques will fail if you are only able to apply them sporadically. *Consistency* is the secret of success.

I intend to clearly define how to achieve and maintain this consistency. The system is simple and painless. However, finding a professional who will work with you may not be as easy as it is to use the system itself.

Are you an unusually sensitive person, one whose moods can change easily? Many people suffer from mood inconsistencies, but not all to the same degree. The measurement scale is a very personal one. You can measure the degree to which you suffer only by your own emotional yardstick. Attempting to compare your yardstick to someone else's is not only frustrating, but fruitless. The only comparisons you can make are how much better you want to feel, and how much more consistent you want to become, according to your own yardstick.

You may try to reason with yourself and attempt to explain away your mood swings. You may say to yourself that there must be a logical reason for them, and may even berate yourself if you don't find it. After all, aren't we all logical and reasonable people? If so, then a rational explanation should be at hand. However, in reality we rarely find one. We may manufacture a reason which makes us temporarily comfortable. But "deep down," where it really counts, are we truly satisfied with the explanation?

Would you really be surprised to find that your frustration, anger, or depression is not because your mate is in a rotten mood, not because your best friend forgot your birthday, not because you forgot to buy milk on the way home and now you have to go out again? There is a deeper reason. That mysterious, hidden cause is what this book is all about. Your inconsistent mood swings are actually caused by a *genetic trait* that you inherited from one or both of your parents, combined with an environmental *mineral deficiency*.

I do not want to create an attitude of helplessness, nor do I want to encourage a feeling that individuals are not responsible for their behavior.

This book offers not only an explanation of behavior, but also a method to modify it so you can be more comfortable with yourself and with the people around you.

More and more genetic research is confirming that what we are depends mostly on what we inherited from our parents, and that the environment we live in is only a tempering factor. Our hair, skin, and eye colorings are obvious hereditary traits, along with our general builds and bone structures. Levels of intelligence and degrees of creativity run in families. The degree of sensitivity to chemical balances (or imbalances) appears to be inherited. Sensitivities to many medications, vitamins, and minerals are now believed to be passed from parents to child. Personality types and emotional characterisitcs are likewise inherited.

High and low emotional levels are inherent in every person's psychological development. These varying levels make us the sensitive, emotional individuals that we are. But when our emotional highs and lows reach such intensity that they interfere with our internal comfort and cause inner turmoil, they are no longer conducive to psychological health. The cause of such emotional turmoil is generally a lithium deficiency. While the deficiency is environmental, the degree of your sensitivity to the Lithium deficiency is inherited. Extreme mood swings are genetically passed on to you in much the same way as the color of your hair.

If you are anemic because you have an iron deficiency, you will remain unable to form adequate supplies of blood no matter how hard you try. I can beg, bribe, or threaten you, but you will still remain anemic unless you receive an external supply of iron.

So, too, if you are deficient in the mineral Lithium, you will remain unable to control your moods. It might almost seem as though you are two people—one happy and free, and the other depressed and suffocated. I could beg, bribe, or threaten you, but it would be to no avail. Without Lithium as a stabilizing agent, you can't seem to find the middle ground or to have consistent reactions to life's many stresses. Even assertiveness training or conventional psychotherapy cannot be effective as long as you are Lithium-deficient.

This book has been written to provide an intellectual understanding of your mood swings, to supply you with enough information to determine whether you may be sensitive to a Lithium deficiency, and to make you aware of a practical solution to overcome your mood swings and find the internal peace that mood consistency can provide.

Author's note: A sensitivity to a Lithium deficiency is found both in males and females. In many instances throughout this book, the male gender pronoun is used to cover the "he/she" situation rather than interrupt the flow of material for the reader.

Chapter One

PEOPLE HAVE COMPUTERS—TWO

The mind can be compared to a complex system of computers. Our senses acquire information and the mind stores it for future reference. Push the right button on the computer console, and the information is retrieved. However, we are a system of two computers. The constant interaction between them may cause incorrect retrieval or manipulation of information.

Each of us is born with two major, but separate, computers. One is the *thinking* computer, commonly referred to as the "brain." The other is the visceral "gut-level" or *emotional* computer, commonly referred to as "the heart." We all use both of our computers, but to different degrees. Most people can be separated into one or the other of two basic personality categories, "thinking" or "feeling" types.

"Thinking" computer personalities are characterized by organized, intellectual, inductive- and deductive-logic-type thinking. They are generally practical by nature. They think in lists and words, and are self-starters and implementers of plans and ideas.

"Feeling" computer personalities are commonly associated with the subconscious mind and the instincts. These individuals basically disregard logical connections, permit contradictions to exist simultaneously, have a here-and-now conception of time, are impulsive, and desire immediate gratification of their wishes or desires. "Feeler" types are also frequently associated with the capacity for creative thinking and thinking in pictures or images.

"Pure thinkers" and "pure feelers" are hard to find. Most of us are a mixture of both. However, the majority of us have many more of the characteristics of a "feeler" than we do of a "thinker." In my opinion, at least 75% of the population is represented by predominantly "feeler"-type personalities. Most of our decisions are made by the emotional computer. At the moment of truth, the "feeling" computer overwhelms the "thinking" computer.

In the past, people who reacted emotionally to situations were considered weak, naive, gullible, and intellectually disadvantaged. On the contrary, the more intelligent people are, the more likely they are to be naive, gullible, and "taken advantage of." These personality traits are for the most part inherited. To view them solely as weaknesses creates an incorrect impression. They are only weaknesses when not kept under conscious control.

On the input console of our "feeling" computer is a large button marked "R"—for Rejection. We all have this button, but for some of us it seems to be more sensitive to even the slightest pressure than it is for others. The impact of rejection and the way it is handled dramatically govern our attitudes, moods, feelings of self-worth, security, methods of interacting with others, and even intellectual capacities. For most of us, the conscious or subconscious fear of rejection is the prime motivating force behind almost all our thoughts, actions, and interactions. The fear of rejection can be either acknowledged or denied, but that only determines how the fear is handled and not whether it exists. We constantly try to avoid feelings of helplessness, insecurity, inferiority, and worthlessness. A "feeler" seeks praise, acknowledgment, and approval from other people, while a "thinker" is more able to obtain it from within. "Thinkers" are most often happy with the personal satisfaction of a job well done, whereas a "feeler" needs to hear and feel the same reinforcement from others.

Many wealthy people who have amassed their fortunes by their educational or business accomplishments have the greatest fear of rejection and criticism. The anxiety they experience over rejection can literally overwhelm them. They feel an all-consuming need to be liked, accepted, and surrounded by people. They are willing to give and give again to avoid loneliness or rejection, to work and work some more to be assured of success. Their business or professional judgment usually remains intact and appears logical. However, often even in this area, the "feeling" computer may prevail over the "thinking" computer.

The structure of your personality is influenced more by genetics than by culture or environment. However, the degree to which your innate potential is realized is modified by environmental and cultural factors. You are born with a certain amount of sensitivity to rejection in your basic personality. However, if you were excessively sensitized to rejection during your formative years by experiencing repeated rejections, your sensitivity to rejection may rise. An individual who was born with greater rejection sensitivity could become even more highly sensitized, and may react even more readily and intensely to any real or perceived rejection.

As we continue to grow and develop as individuals, our rejection sensitivities can be changed by new experiences. Many positive experiences later in life can overcome negative early encounters. However, early memories (known as engrams) are the most intense and most difficult to change or eradicate. Even though we may not be consciously aware of these memories, they continue to exert a subtle influence. People who have had very positive self-images are able to withstand a great deal of stress in later life. Those who have developed negative self-images are at a much greater risk when encountering negative experiences later in life.

It is important to point out that rejections may be either real or perceived. What one person views as a rejection, another person may not. For example, some people might regard you as rejecting them if you were to pass the rolls to someone on the right at a dinner party, rather than passing the rolls to them on the left. Such a "rejection" would only be obvious to the individual feeling slighted. In other situations, the rejection is much more obvious, such as when a mate finds a new lover and asks for a separation or divorce.

People are born with greater or lesser degrees of sensitivity in all personality areas, rejection sensitivity being only one. Life's experiences can vary the degree of sensitivity, but they do not play as great a role as the genetic endowment.

Even the methods individuals use for problem solving are

inborn. At one end of the scale is the obsessive, represented by an individual who reviews the same information over and over again. Those at the other end of the scale prefer not to answer or make decisions. They either make no decision, or they let others do it for them.

Most of us are somewhere in between these two extremes. A "thinking" type is generally an obsessive problem solver. "Thinkers" tend to organize and reorganize their thoughts, to make lists, and to think over all the possibilities and the effects of possible solutions. "Feelers," on the other hand, tend to handle problems by avoiding them. They develop the "ostrich syndrome": they quickly bury their heads in the sand and hope that when they pull their heads out, the problem will have either gone away or been solved by someone else.

Individuals who are primarily the "feeling" type are emotional, sensitive people. They are the kissy, huggy, touchy ones. They are daydreamers. Some daydream to such a degree that while driving, they may suddenly find themselves three or four miles down the road and then shake their heads and wonder how they got there (a situation called "highway hypnosis").

A "feeling" person, one who has a relatively high emotional sensitivity, can usually achieve a deep hypnotic state. Unfortunately, hypnosis has rather frightening connotations for many people. This is mostly due to the misleading portrayals of hypnosis shown on television and in the movies. The hypnotic state is actually like daydreaming. It is self-motivated and, for the most part, under the control of the individual entering the trance. In fact, all hypnosis is really self-hypnosis. The ability to enter a trance is a genetic trait. The gene which controls the depth of trance capacity is inherited in parallel with the gene for emotional sensitivity.

The more of a daydreamer you are, the more your personality tends to be of the visceral or "feeling" computer type, and the higher your sensitivity to a Lithium deficiency appears to be. You are more emotional, more sensitive, more giving, and you can readily "feel" and understand others' emotional states.

"Feelers" tend to share still other specific personality traits. For instance, they are excitable, impulsive, enthusiastic, impetuous. They are "here-and-now" people. They lean toward belief in ESP, astrology, and psychic phenomena, even if only by giving them subconscious credulity. "Feelers" are usually the ones trying to gain a deeper understanding of psychic healing, clairvoyance, telepathy, precognition, and especially deja vu."Feeler" types have been given many genetic gifts of intuition and subconscious understanding. However, the most gifted ones are often also genetically counterbalanced with feelings of insecurity, helplessness, worthlessness, and despair. It is sometimes difficult for highly intuitive people to trust their intuitions. This interesting paradox of nature typifies Lithium-sensitive individuals.

Usually a "feeler" is one who is sensitive, loving, and giving. However, some "feelers" may not be as responsive as others. They may have been sensitized by rejection or negative criticism to such an extreme that the "feeling" computer was overwhelmed and the computer's "rejection module" was short-circuited. The result is a non-feeling "feeler." The pain of feeling has become too exquisite.

"Thinkers" may also be very loving and giving. A "feeler" hugs and kisses; the giving is very personal. A "thinker" may instead show his love by giving material things. His affectionate gift is no less meaningful, just different.

Throughout this book, I will continue to identify the personality characteristics of the two types of "computers" and their variations. Most people have traits of both the "feeler" and the "thinker," but to differing degrees. You will readily see that the majority are "feelers." Most importantly, you will recognize that most of us are inconsistent in our moods, and that the cause is a genetic sensitivity to a lithium deficiency.

The effectiveness of every type of human relationship — husband-wife, parent-child, employer-employee, etc.—depends on consistency. For the "feelers" who find it difficult to remain consistent, therefore, the use of small amounts of Lithium Carbonate (a safe, easily dispensed mineral supplement) may be in order. Lithium Carbonate can produce an internal environment of peace and tranquility, a state which many of us have sought for most of our lives.

Chapter Two

THINKERS ARE FEELERS, TOO

Because of the variations in personality styles of "feelers" and "thinkers" there can be multiple combinations of personalities who marry. Marriage is one of the most closely interwoven interpersonal relationships. It usually presents a complex variety of interactions. A "thinker" and a "feeler" may experience a great deal of difficulty in understanding each other's needs and in satisfying them.

Anne is a 29-year-old white female. She is a high school graduate, and prior to meeting her husband, Mike, she worked in an office as a secretary. Mike is a 35-year-old white male. He is a college graduate who works for a major company in a middle management position.

During the early part of their marriage, their social and intimate relationship was very intense, spontaneous, and rewarding.

Mike grew up in a work-oriented, male dominated family. His overall orientation prevents his wife from having the freedom of spontaneity in a friendship situation. If Anne expresses a self-actuated opinion, Mike views it as an intentional prelude to marital discord.

Anne, on the other hand, is basicially a compliant individual who spontaneously gives, loves, and cares, and when left to her own devices, is able to fulfill most of her husband's needs. However, when demands change what once was a voluntary offering into a painful obligation, the spontaneity disappears and the interaction becomes negative.

When they were married about five years, Anne developed a severe migraine symptom. Because of acute onset and intensity, a full medical and neurological evaluation was needed. No physical basis was found to explain her pain, although she suffered from this syndrome for many months. Consequently, Anne was referred for psychiatric evaluation and treatment.

The psychiatric evaluation uncovered that the negative cycle of their interaction became obvious and explosive when Anne was no longer comfortable relating to her husband in the manner to which he had become accustomed. Most of the changes in her feelings, however, were subconscious. Anne viewed her sexual drive and her desire to please Mike as a natural continuation or outgrowth of the day's events. If Mike came home from work in an angry, hostile mood, he was likely to be silent and not respond to Anne's attention, causing her to feel rejected. As the evening progressed, her mood deteriorated as his improved. Nonetheless, he still expected her to be a spontaneous, loving partner, able to interact with him sexually. He felt that it would bring about a positive change in his own behavior. It was very difficult for Mike to understand why Anne was unable and/or unwilling to relate to him in a prolific, varied, and adventuresome sexual interaction independent and irrespective of the day's events.

For several months prior to the decline in their sexual interaction, Mike had been giving Anne the "silent treatment" if she disagreed with him about their relationship, their son, how to spend money, their home, in-laws, etc. He remained silent from hours to days at a time. Initially Anne reacted to these silent periods by asking, cajoling, imploring, and finally pleading with him to speak. She prepared his favorite meals, wore the clothes he liked best, etc., but most times it made no difference. She could not understand the reasons for his new behavior and she felt overwhelmed.

Mike is a hard working, industrious man who is devoted to his family. His wife and child are his number one priority, and he wants to give them some of the finer things in life. Although he enrolled the family in a summer health and swim club, he resented Anne going to use the facilities. For several summers, he'd come home from work on Friday, tired or exhausted, but he refused to go to the pool with her on Saturday. Instead, Mike spent his time working laboriously in the garden and resented Anne going to the swim club and not staying home with him to help with the garden. She told him she'd garden later in the day when it was cooler and again implored him to go to the swim club, but he wouldn't change his plans. He felt rejected, unloved, and uncared for. He would then build up a high degree of anger, frustration, and contempt for his wife. When she returned from the club, he became argumentative, negative, and belligerent, instead of expressing his true feelings—that he missed her and felt hurt. The weekend would once again be ruined.

If she went to the swim club during the week, he resented the people she spent time with, and was annoyed that the house was not impeccable, though it must be noted that the house was almost never dirty. It was, on occasion, scattered with toys with which his young son and friends had been playing.

Anne continued to become more and more dismayed and confused. She reminded him that he had told her he wanted her to use the club as much as possible, but that when she used the facilities on a routine basis, he became angry, frustrated, and accusatory. Anne is psychologically orientated toward simplicity and directness: when Mike told her to use the club as much as possible, she responded accordingly. It was difficult for her to deal with the fact that what her husband said was not what he meant, and what he meant was not what he said. Mike *wanted* Anne to use the club *and* stay home with him at the same time. He never clarified his statements, nor would he admit that they were in conflict.

Mike could not admit to himself and/or deal with his own emotional sensitivity and fear of rejection. He interpreted any perceived affront as an overwhelming attack on himself as a person and on everything he represented. He felt Anne should have been able to readily deal with nonspecific, double messages, and cater to his verbal and nonverbal needs and whims.

The swim club was only one example of miscommunication. Over time, misunderstandings and misreadings of each other's behavior and personality styles increased and became almost continuous.

Anne is a "feeler." She is an individual who does not readily use inductive or deductive logic, but responds instead to the primary process intuitive system for problem solving and decision making. Mike, on the other hand, appeared to be a "thinker." He is very cognitive, intense, and logical in his decision making processes. It was extremely difficult for Mike to *understand* how Anne *felt*. This paradox in itself is the root cause of many marital problems. Mike and Anne's situation is further complicated by the fact that Mike acts like a "thinker" toward others, but hurts like a "feeler" within himself. He appears in control externally, yet internally, he is very emotional.

We all must recognize that feelings are incongruous and unreasonable. They cannot be substantiated, and are contradictory as well as illogical. When you feel bad, no amount of logic is going to make you feel good.

As noted earlier, Mike is a middle-level executive for a well-known firm. He practices his profession in a very austere, businesslike, matter-of-fact manner. Most people who are involved in the business world have a "split personality." They present one personality or personality constellation at work, and a second one at home. The way they appear at work is not necessarily the way they actually are. We must recognize that the business world personality may be significantly different from the at home personality

There are many opportunities for the person at work to blow off steam. There are also many chances for personal contact and growth. The spouse who does not work outside the home, however, has limited outlets and contacts, limited ability to express feelings, and fewer opportunities for personal growth. Consequently, extended periods of time may be spent without exposure to other human vibrations and input. People who have a great need for these vibrations and input must then rely on substitutes such as the telephone and television. Housewives are sometimes belittled because of their attentiveness to afternoon soap operas. The soaps are, however, better than no human contact at all.

Because the working member of the family has far more opportunities for expression of and contact with feelings, you can readily see why there is often a sense of desperation on the part of the wife when the working spouse returns home. There is often a need to discuss the affairs of the day immediately, and to have an embrace or other sign of recognition, especially if the homebound spouse is a "feeler" type.

The male counterpart, on the other hand, needs space or distance when he gets home, to relieve himself of the stresses and strains of his day. He may be more interested in a casual or perfunctory greeting, and then seek out his own space. The homebound spouse may feel hurt, rejected, and abandoned. Most times she overreacts to the immediate circumstances at hand. Reality and rational thinking rarely apply in these situations.

The male now feels justifiably annoyed with his spouse for placing additional demands on him after he spent much of his energy getting through his busy day at work.

On the other hand, if the working spouse has had a relatively good day at work, and only a minimal draining of his reserves has occurred, he may greet his spouse very warmly, pleasantly, and in a manner she considers more appropriate. Although it would seem this type of greeting would be more welcome, this in itself leads to ongoing confusion and misunderstanding because of the inconsistency. We recognize that a moderate amound of basic inconsistency is not only expected, but useful in an ongoing relationship. It is the gross, frequent, and dramatic inconsistencies that are destructive to a relationship because they lead to misunderstandings, second–guessing, and hurt feelings.

As with everything else, there are exceptions to this situation. There is the individual who is extremely sensitive and emotional and who, despite his work, still interrelates with his spouse effectively, efficiently, and frequently.

It has been shown through the years that inconsistent behavior creates anxiety, tension, and confusion. The homebound spouse, although subjected to a variety of emotions because of her environment and its demands on her, is expected to understand and forgive the working spouse. She is expected to place her own needs and desires aside, while devoting herself to her spouse and all of his needs. The inconsistency in the greeting process causes overwhelming, chronic anxiety, a disruptive and painful situation. It is our first reference to the ongoing cycle of anticipation and disappointment, a cycle which most of us must deal with on a daily basis.

Rarely does an individual evaluate someone else's behavior as an isolated entity. Instead, it is basically evaluated in light of a preset anticipation of how one would like the other individual to behave or react. This degree of anticipation almost automatically breeds disappointment. Many relationships have been essentially destroyed because anticipated behavior did not occur, rather than because of something that actually happened. The anticipation of the behavior and the subject of disappointments are probably characteristic of a "feeler" computer type. That person's desperate need for another human vibration leads to vulnerability and ongoing emotional pain. It is very important to deal with this concept of ongoing anticipation and disappointment, or "antisappointment."

An individual who has the constellation of characteristics we are defining is one who desperately and concretely anticipates the behavior of another, usually based on a model of what his or her own behavior would be. The anticipation is so intense and specific that when the other individual does not respond accordingly, there is great emotional pain and suffering. From a clinical point of view, the following addendum to the golden rule is one that every individual with this psychological makeup should adopt. "Do unto others as you would have others do unto you," is a proverb which the "feeler" individual adheres to, even to his own detriment. For that individual, the golden rule would be more useful if it read: "Do unto others as you would have others do unto you, *but don't expect them to.*" It is the expectation of another's behavior that causes great pain and almost automatically builds disappointment into the system. It is self-defeating, nonproductive, and yet repetitive.

A "feeler" is an individual whose time frame is in the present, who does not learn well from the past, nor project far into the future. However, when it comes to another person's behavior, there is a projection into the future, although the time frame for this projection is, in itself, limited. An individual with this personality constellation is an individual who tends to repeat history. Despite the recurrent disappointments which are suffered as a result of someone else's anticipated behavior, the "feeler" continues to anticipate others' behavior according to his or her own needs and desires. You would think that one would learn from painful experiences, but experience is, at best, a poor teacher for the "feeler." It becomes distressingly obvious that the greater an individual's anticipation, the more surely he has painted himself into a corner. Options become extremely limited and the sense of being overwhelmed, even victimized, becomes apparent.

Anne and Mike continued to suffer for a considerable period of time. She continued to believe she was responding to her husband's wishes, but was becoming more and more anxious and confused because she was attempting to simultaneously deal with both his verbal and nonverbal desires. At the same time, Mike was unable and/or unwilling to deal with his own multiple or complex needs. Externally, he continued to present himself as a macho type. Either consciously or subconsciously, he refused to recognize his overwhelming need for recognition, his fear of rejection, and his intense overreaction to criticism. Simultaneously, he continued to place inordinant, unreasonable, and increasingly intense demands on his wife. Anne became increasingly frustrated and overwhelmed by her inability to please her husband. His immediate needs and desires were never clear to her.

Anne found herself in the position of being in a "double bind"—being "damned if you do and damned if you don't." She became confused in her attempts to deal with his conscious demands on a conscious level, while at the same time she wanted to deal with his subconscious demands, which were the ones she felt he really wanted her to meet.

Returning to the example of the swim club: on a conscious level, he wanted her to avail herself of the facilities so that both she and their son could enjoy themselves. On a subconscious level, however, he wanted her to be home, to have an impeccable house, and a full meal waiting for him upon his return from work. This situation became increasingly stressful and created marital discord.

The therapeutic aim for Mike was to have him develop the

skills for dealing with his own needs and desires, while simultaneously helping him learn to understand how Anne "feels." Anne needed to get a better appreciation or understanding of how Mike both "thinks" and "feels." This particular type of therapeutic interaction is frustrating at best. We are not only dealing with multiple personalities, but with multiple levels of interaction simultaneously.

I spent inordinate amounts of time citing examples to Mike and Anne to clarify the situation and to get them to recognize what they both had to do in order to facilitate communication. My clinical experience has convinced me that proper communication techniques are the foundation of all relationships, and depend on a full understanding and appreciation of both the basic "feeler" and the basic "thinker" computer types. However, consistency remains the most crucial factor. In order to develop a sense of trust within yourself and a trust of others, you must maintain an internal environment of comfortable consistency. Only with this consistency within yourself can you appropriately deal with the needs of others. Concurrently, you must expect and obtain relative consistency from those individuals with whom you deal, especially from the individuals with whom you deal most often. Without this element of consistency, the communication techniques of therapy itself would be undermined. It is not only *what* the therapist says that is most important, but also *how* it is heard by the individual or individuals.

We came to a therapeutic impasse in attempting to implement these communication techniques with Mike and Anne. I could not easily penetrate Mike's unwillingness to accept the fact that Anne had as much right to "feel" as he had to "think." I tried to have Mike understand that his inability or unwillingness to deal with Anne's "feeling" computer was as much an emotional issue as Anne's "feeling" computer itself. Mike's refusal was an emotionally based concept. In truth, his stand was just as irrational and illogical as he thought Anne's basic personality style to be. What Mike was unwilling to recognize was that his internal environment was that of a "feeler," not a "thinker." He refused to recognize the inconsistencies in his overall behavior and his overall thinking patterns. Actually,

Anne was the more consistent of the two. The way she felt internally was the way she presented herself externally. Her pattern was relatively less complicated and much more apparent.

Mike, on the other hand, presented an external picture which was incongruous with his actual internal emotional environment. It was his own internal inconsistency which he refused, or was at least unwilling, to recognize. Consequently, in order to deal with Mike and Anne as a unit, we had to get Mike to understand his need for consistency, and then to help him effect it.

At this point in therapy the use of Lithium Carbonate as a mediator of consistency was introduced. I was prevented from introducing this most necessary mineral earlier on in therapy because I knew how Mike felt about medication and psychiatry in general. I first had to establish an intense, therapeutic alliance. I had to be sure that presenting the necessity of the mineral to Mike would not cause him to abandon therapy. Fortunately, the therapeutic relationship was of sufficient quality to allow me to enter into this arena with Mike. I had to continually reassure him that Lithium is a mineral and that it is a naturally occurring substance in nature. I had to stress that without Lithium, the consistency necessary for a positive therapeutic outcome was not within our grasp.

Our timing was augmented by the fact that Anne was rapidly reaching the limit of tolerance of Mike's behavior, which often happens in therapy. She could no longer deal with his explosiveness, his argumentative nature, and his demanding ways, but most of all, she could not handle his temper tantrums. The threat of her leaving him allowed him to overcome any resistance to the use of Lithium which still remained.

Mike grudgingly began his course of mineral replacement. His initial level of cooperation was, at best, suspect. Fortunately, blood tests are available to check an individual's Lithium level. Therefore, we could measure his degree of compliance to find out whether or not he was taking the prescribed dosage.

Much to his surprise, Mike found that the frequency and intensity of his temper tantrums decreased markedly. Conversely, much to her "antisappointment," Anne found he was not doing as well as she had anticipated. They each had a different idea of what the Lithium therapy would initially accomplish. Mike expected it to do nothing, and was pleasantly surprised. Anne expected it to be a miracle cure and was disappointed. I had to explain that patience was a factor to be considered and that, in time, the results would be very positive and concrete. Mike continued his prescribed mineral replacement with more enthusiasm, and Anne patiently watched his progress.

Mike and Anne are doing much better now. They enjoy each other's company. They are able to share feelings and discuss their weaknesses as well as their strengths. They are even considering enlarging their family.

Chapter Three

THE LITHIUM HOAX

In a lecture I recently attended, one of the main speakers defended a myth as "something that never was true and always will be." Many myths have existed about the mineral Lithium. Lithium has been presented as a drug that is difficult to administer and that must be carefully and stringently monitored. The most misleading myth is that it is used only by psychiatrists for the most seriously, emotionally ill people.

The fact is that Lithium Carbonate is a safe, and readily usable medication. It is not a man-made drug, but a naturally occurring mineral that everyone needs in order for the mind to function properly and cohesively. Lithium's usefulness is widespread, dramatic, and essential to a person's well being. It is easy to use and has a wide range of applications. I cannot stress too emphatically the ease, appropriateness, and almost universal application of this mineral. Individuals with a sensitivity to Lithium deficiency will have mood swings. Their intensity and severity are in direct relation to the severity of the deficiency and the degree of sensitivity to it. The sensitivity to the deficiency is hereditary; the deficiency is environmental.

When Alcatraz Island was established as a federal prison, a hoax was perpetrated on the first inmates. They were told by guards that there were man–eating sharks in San Francisco Bay. From time to time, something resembling a shark fin could be seen in the water, lending added credence to the guards' statements. As additional inmates were put in the prison on Alcatraz Island, the first group of inmates perpetuated the myth. As inmates continued to relate the story to other new inmates, there was no longer any question about whether or not there were sharks in San Francisco Bay. This "fact" became a self–evident truth.

Likewise the stories and misconceptions about the use of Lithium Carbonate as a therapeutic tool have been perpetuated. I would certainly acknowledge the dangers in attempting to escape from Alcatraz by swimming across San Francisco Bay, but the existence of sharks is not one of them.

There are inherent dangers and difficulties in using any medication. Most of the dangers and difficulties ascribed to Lithium Carbonate are as legitimate as the sharks in San Francisco Bay. The Lithium Hoax, like the Alcatraz hoax, has become a self–perpetuated example of misinformation. Stories somehow become embellished as they go from one person or source to another. As the information about Lithium Carbonate was disseminated, it fell victim to misunderstanding, exaggeration, and misapplication. It was thought to be the province of psychiatry rather than a province of all fields of medicine. To further complicate this already complex situation, most practitioners in the mental health field are unfamiliar with the appropriate application of Lithium Carbonate. I believe Lithium Carbonate can be used successfully by the vast majority of practicing physicians. Once the ease of use is clearly explained and understood, Lithium Carbonate can be prescribed and utilized extensively by general practitioners of medicine and surgery. These doctors still remain the first line of medical practice and should be taught how to utilize and prescribe Lithium Carbonate to help the many individuals who suffer from the emotional stress that results from a sensitivity to a Lithium deficiency.

There is an analogy to diabetes. The average general practitioner is well able to care for most diabetic patients. They refer the complicated diabetic to a specialist because he has additional training. Likewise, the average general practitioner is certainly capable of handling most Lithium sensitive individuals. The complicated cases, or nonresponders, should see the specialist, a psychiatrist versed in the application of Lithium Carbonate. There is much controversy within the field of psychiatry itself about the use of Lithium Carbonate. The most vocal nonacceptors have had the least exposure to the positive results of Lithium Carbonate therapy.

It is important to understand why Lithium Carbonate must

be given as a mineral replacement and what has happened to our natural supply.

Since the start of the Industrial Revolution ninety years ago, there has been an ever–changing supply of natural minerals and vitamins and an ever increasing pollution of our waterways and our air. Because our streams are polluted, we have begun to filter and refine our water supply before it comes into our homes and businesses, and Lithium Carbonate has been almost totally filtered out of our natural water supply. It is a mineral found in only a few natural food sources, and then only in minute amounts. Because of the decreased supply, there is a natural deficiency.

This is like an ancient mariner aboard ship who has been deprived of fresh fruit. Without fresh fruit there is a lack of vitamin C which leads to scurvy. To treat scurvy, vitamin C, in the form of fresh fruits, is given. To alleviate Lithium deficiency, you must take an external supply of Lithium Carbonate. Without ample supplies of this natural mineral, you will be likely to develop a multitude of personality changes, such as moodiness and a circular–type mood disorder. In its earliest stages, this is called a cyclothymic personality disorder (or cyclical changes in an individual's moods, but to a mild degree). In its most advanced form, it is called a bipolar mood disorder (or manic–depressive illness), bipolar meaning two poles or excessive highs and excessive lows. A person cannot control these excesses. They are usually intense and violent. They may vary from one to the other, or they may be recurrent highs or recurrent lows. There may be a preponderance of one or the other extreme. The treatment is to replace the mineral. Thus, the concept exists that this moodiness is internal and without obvious or reasonable external cause. The mood changes can run the spectrum of feeling unusually pressured by life's daily stresses to severe emotional illness. Some of the most prominent features of Lithium deficiency are moodiness, explosiveness, insecurity, inferiority, hostility, and, most of all, a general feeling of not being in control of your own moods.

In each successive generation since the Industrial Revolution, the amount of Lithium Carbonate has been increasingly depleted. The aforementioned symptoms and/or personality traits have become increasingly more apparent and more destructive with each generation. Impatience, intolerance, and the unwillingness or inability to be able to listen to someone else's viewpoint, as well as frequent feelings of helplessness, hopelessness, and despair have become more and more pronounced.

It is quite possible that the significant increase in the divorce rate, crimes of violence, and perhaps even the excessive use of alcohol and mind altering drugs are at least in part traceable to Lithium deficiency. Each generation over the past several decades has resorted to greater and greater reliance on an external means to achieve peace of mind. The rate of drug use has reached epidemic proportions. Clinically, I have seen a large number of people spontaneously discontinue the abuse of drugs when taking Lithium Carbonate, even without addressing the issue. Binge or cyclical drug and alcohol abusers appear to respond even better to Lithium Carbonate therapy.

When an individual is placed on Lithium Carbonate therapy, there is often a feeling of negativism about the usefulness of the mineral. It is necessary to stress repeatedly that Lithium is a mineral and *not a drug*. The only similarities it shares with classic drugs are the fact that it is purchased in a drugstore and that it is a natural "chemical."

It is useful to compare the mood disorder of Lithium deficiency to the metabolic disorder of insulin deficiency. An individual who is diabetic has an insufficient internal supply of insulin and needs an external supply on a daily basis to maintain the internal environment with reference to sugar metabolism. An individual with a mood disorder, who requires Lithium to stabilize, needs a daily external supply of Lithium Carbonate. Lithium therapy is that simple. The complicated part is getting people to recognize how simple it really is.

Most people in my practice who have been placed on Lithium Carbonate are taking subclinical doses. By subclinical, I mean that most of these people take only two or three capsules of

Lithium per day. This minimal, daily, external supply is usually sufficient to stabilize the moods of most people and since the overall amount is minimal, there is a minimal need for blood tests and observation.

An individual who suffers from this mood disorder often slips into an altered mood without an obvious explanation. A person who is in a good mood may become depressed. If someone asks, "What happened? You were okay a minute ago and now you're down in the dumps. What's wrong?", the individual may answer, "I don't know." The person is then challenged: "What do you mean, you don't know? How can this happen and you not know? What is the matter with you?" The individual then responds, "I really don't know."

The momentum of the conversation and the tempers usually begin to rise. The individual is challenged once again. "You must know, and you won't tell me, so that means you don't care about me or trust me." The individual may then make up an excuse or reason to pacify the questioner. If the excuse later proves false, the individual is accused of lying, when the only real problem was not being able to explain the mood change. The individual then feels misunderstood, mistrusted, misguided, and foolish. The truth is that in most instances of mood changes, the individual does not know what happened. This uncertainty leads to feelings of helplessness, hopelessness, inferiority, and insecurity. This spontaneous change in mood is what psychiatry calls an endogenous (from within) mood change or mood swing.

As these episodes recur, the stability of the relationship between the individual suffering from the mood swing and loved ones becomes unstable, at best. Disappointments and arguments flourish; mutual distrust becomes intense. The very nature of this illness is that these mood changes most often occur irrespective of what is happening around the individual. If an incident in the external environment does precipitate a reaction, the individual overreacts. The longevity and intensity of the reaction is usually out of proportion to the particular circumstances.

The situation is compounded even further when the loved

one avoids the individual because of his own frustration. The overall relationship suffers yet another hardship. The lack of contact causes the individual suffering from mood swings to become worse, and much more frightened and insecure. This arises from feelings of rejection and isolation, so the individual with the problem continues to give answers and responses he thinks the other person asks for or requires to prevent the feeling of loneliness and to avoid feeling rejected.

The situation is compounded again because the at-risk individual recognizes that the answers he is giving may be untrue. Yet, he is sure the answers are necessary to maintain an already unstable relationship. In the classic sense, this may be considered a "lie." The person does not believe that it is a lie, but thinks of it as offering a gift to maintain the relationship. His "computers" have been overwhelmed, confused, and he has become desperate. Above all else, he feels he must avoid the feelings of rejection and isolation.

This situation becomes increasingly tense and complicated. The individual at risk creates a story as a gift which in itself is difficult. It is then compounded dramatically when the person begins to vary the explanation according to what he believes the other person wants to hear.

If George is building a new house, when he talks to his friend Tom, he may emphasize the siding because Tom likes siding. When he talks to Bill, they discuss plumbing; with John, it's brickwork. George adjusted his interest to suit the primary interests of his friends.

Each of George's friends now believes that his particular interest is George's primary one, until they all get together. Then, they may accuse George of lying to them.

When the "feeling" computer person is caught in this situation, the individual will vary the story and tailor it to everyone who requires the response, since the "feeler" is highly intuitive and sensitive to other people's needs. We then have a situation where eight different people might receive eight variations of the same story. The story may or may not be in any way truthful, but, even if it is, it is embellished according to what the storyteller "feels" the other individuals need to hear. The situation becomes overwhelmingly distorted, and the consequences can be dramatic and painful because the individual was unable to consider the consequences when the answer was given.

The at-risk person creates a story because the listener or questioner does not believe the truth (that the individual suffering from this mood disorder does not "know" why there has been a change in mood). The listener becomes convinced that the individual who is suffering from mood swings did not tell the truth at the start. It has been said that a lie told immediately is often more believable than the truth. This situation is obviously one that applies to the person who suffers from a circular mood disorder.

Lithium offers the individual stability of mood. Without mood stability and consistent reactions, you cannot be "trusted." You cannot trust yourself, and others cannot trust you, either. (Without consistency there can be no trust, and without trust there can be no love.) Lithium offers stability of mood for the emotionally sensitive individual. The extreme peaks or "highs" and the extreme depths or "lows" of depression are for the most part controlled significantly. Perhaps for the first time in his life, the individual is in control of his moods, instead of his moods controlling him. The extreme peaks and valleys are removed, but the individual is not "mind controlled." The individual does not become an automaton or robot. Moods vary by the situations that present themselves. Overreactions and misguided, inappropriate, severe, and destructive intensity of reactions are brought under control. The individual is able to act rather than react and is able to think about a response rather than responding spontaneously.

The average individual suffering from mood swings reacts rapidly and spontaneously. A mood swing type individual is so rapid in the thought processes that the response to a question is often stated without conscious thought. The subconscious mind deals with problem solving and the solution "pops" into the head.

Lithium Carbonate allows individuals freedom of choice. It allows individuals who suffer from a circular mood disorder to act rather than react. It provides a delay factor which will let the brain be put in gear before the mouth is put into motion. Lithium allows a person to think and consider the consequences before the thought is expressed or the action committed.

When Lithium deficient individuals first begin to take Lithium, they report most frequently that they "begin to slow down and calm down." They often expressed a sense of quiescence. They no longer feel like a forty-five speed record running at seventy-eight (their highs) or at thirty-three-and-a-third (their lows), or even like a series of seventy-eight speed records playing simultaneously. When individuals who have mood swings are in a "high," they often feel as if they are multiple records because their minds skip from thought to thought so rapidly that their conscious minds cannot keep up.

As they continue to take Lithium, they recognize that their memory improves and that they are able to attend to what is going on around them. Their mind is not spinning so quickly that they are unable to deal with the environment nor are they so depressed that most of their energy is being drained from them.

Memory consists of many factors. When one's memory is called upon, one is asked to recall an event or thought. Before a thought can be recalled it must be retained, and before it can be retained, it must be registered in the mind. People suffering from mood swings have a great deal of difficulty registering memories because of their inability to attend to the external environment. Lithium helps a sense of calmness to set in, and events and comments from other people are now more readily placed in perspective. This perspective is one of reality. One can deal with information in a more appropriate manner, and can deal with a situation as it exists.

Lithium offers consistency as well as control. There is a sense of well–being, that one's future will be brighter and more secure. A feeling of self–confidence and self–worth becomes apparent. There is a sense of order to one's life, an overwhelming thankfulness that Lithium has been prescribed.

Chapter Four

THE GENES HAVE IT—WE SHOW IT!

Individuals who are Lithium deficient (for our purposes used interchangeable with Lithium sensitive) are subject to multiple mood swings. These mood swings are often out of keeping with their external environment and the name endogenous (from within) is applied to these swings. At times, however, these mood swings seem to be compatible with the external environment. Lithium sensitive individuals usually react in a manner which is out of keeping with the environmental stress. Their reactions are usually too intense and, at times, overwhelming. When Lithium sensitive individuals react to the external environment, the reaction is so intense as to be all encompassing.

Classic psychiatric literature discusses the mood cycles of Lithium sensitive individuals as major mood swings. Visualize a sine wave curve with its series of ups and downs. These major mood swings, or cycles, are noticeable and can be documented by learning an individual's history. Within these major mood swings lies a series of minor, or more temporary swings. Mood is defined as an overall feeling tone over time. Affect is described as an immediate emotion. For an individual whose overall mood is positive, a temporary negative situation can cause a temporary, short–lived depressive affect. The converse is also true.

It can take days, months, or years for the major mood swings of Lithium sensitive individuals to complete a cycle. If I attempted to diagnose Lithium deficiency based only on these major mood swings, I would overlook most people who are Lithium deficient. The need to replace Lithium Carbonate should also be based on the individual's description of moment–to–moment moodiness.

It is important to take a careful, though brief, history. If I ask specific questions of Lithium sensitive individuals, I can quickly get adequate information about whether or not they need Lithium therapy. This is usually accomplished during the first psychiatric interview. When I directly question individuals about whether or not they consider themselves emotionally sensitive, or whether or not others view them as moody, that is one of the most important, if not the most important, diagnostic criterion of Lithium deficiency.

You must remember that when individuals are in a psychiatric interview, their guard is up. People may present themselves as they want to be seen. If individuals are Lithium deficient, they may also attempt to present themselves as they *feel* you want them to be. Lithium deficient individuals usually have a need to please the examiner. If one is not aware of this, it tends to complicate the psychiatric interview. Historical data is often more useful and more valid than the interview alone in concluding whether or not a particular individual is Lithium deficient.

Lithium sensitive individuals have many similar personality characteristics. As you continue reading, you may come upon descriptions that fit your own personality. This is not surprising because most people are Lithium deficient. These characteristics are so predictable that my approach via communication technique and treatment methods is readily available, so that in a relatively short time people in emotional distress can have their distress alleviated.

The following characteristics are not presented in order of importance. They are presented so their impact can be modified and so they can be dealt with simultaneously by our visceral and cognitive computer.

Court of Advisors

In the old days, a king's court of advisors consisted of a group of people who stated their opinions at his request. Each opinion was fed into a system for the decision-making process. The adviser spoke in the area of his expertise and offered this information for the king's use. The king then utilized this information to arrive at a decision.

To individuals who suffer from frequent mood swings, every part of them and every opinion is an absolute. When Lithium deficient individuals state an opinion, it is not just an opinion but a personal part of them. If this advice or opinion is not followed in its entirety, the situation becomes traumatic. They feel rejected, which leads to hurt and dismay. When someone does not follow the advice of Lithium deficient individuals, it becomes a personal insult and a total personal attack. This might be carried to the extreme and they may begin to feel as if their opinion was requested so that it could purposely be rejected. This may lead to significant psychiatric problems.

Because individuals who suffer from a Lithium deficiency are inherently insecure and uncertain they tend to ask the advice of many different people. This situation dramatically deteriorates if three or four people offer three or four different opinions. Lithium deficient individuals whose basic need to keep loved ones and associates happy are now in an overwhelming quandary. The situation is so intense that they are now literally unable to make a decision. They are unwilling to hurt anyone, so they become even more paralyzed and afraid to take any kind of action. They avoid making a final decision unless absolutely forced to do so.

When one person asks someone else's advice and the person whose advice is requested also suffers from a Lithium deficiency, that is a no-win situation.

Linda, a Lithium deficient individual, was having problems with her monthly menstrual cycle. She went to her doctor who told her she should probably have a hysterectomy. Otherwise, she would have to learn to live with the problems. Before she fully absorbed the implications, Linda found herself tentatively scheduled for surgery in two weeks. Her doctor told her to call him if she decided against surgery.

Linda should have called another doctor to get a second opinion, but instead, called her mother who told her that she was the same age as Linda when she needed a hysterectomy. Linda felt somewhat more assured that she needed the surgery.

Linda then called her best friend, Barbara. Barbara suggested that Linda go to her doctor for a second opinion because she just read an article that stated most hysterectomies are unnecessary. Linda didn't know Barbara's doctor and didn't want to see a stranger about her situation. She also didn't want to hurt Barbara's feelings, so she stalled for time by saying she'd think about it.

Her final effort in looking for help in making her decision was to call her sister, Cheryl, also a Lithium deficient individual. Cheryl told her in no uncertain terms not to have the surgery. She said most surgeries are unnecessary, that there is always a risk of dying while under the anesthesia, and that once the procedure is done, it can never be undone.

Linda didn't want to hurt anyone's feelings (including her doctor's), but she still didn't know if she should have the surgery.

When an insecure individual asks another insecure individual for an opinion, the opinion is rendered in a dogmatic, absolute fashion, usually without thought or prior knowledge about the facts or the entire situation. This opinion is nonetheless delivered intensely and dogmatically. The intensity of the statement covers up in part for the insecurities of the individual whose opinion was requested. The individual who asks the question now has even more difficulty making a decision. The individual who gives the dogmatic opinion, in the usual manner of a Lithium deficient personality, then becomes very expansive and persuasive about his opinion. This may vary greatly from other opinions the questioner recently asked for and with which he agreed. Lithium deficient individuals are placed in a position where they have agreed with many opinions and agreed to follow each of these at various times, even though they often are as different as night and day.

Avoid Rejection at all Costs—Well, Almost . . .

Individuals who are Lithium deficient have an overwhelming fear of rejection and are afraid to be alone. Consequently, they have difficulty dealing with death. That is the ultimate rejection. They become very angry with the individual who has died. Because of the custom set by society, they are unwilling and/or unable to express the anger and frustration they feel toward the individual who has left them, and, keep these feelings to themselves.

The greatest personal rejection is suicide. If people feel they can no longer deal with an experience that becomes overwhelmingly intense, it may be countered by self-rejection and/or suicide. If a business deal in which they have placed a great deal of effort and personal pride should go awry, suicide is a real possibility. The impact of the failure increases if the individual in question has encouraged friends to invest in the venture. The promises to them are echoes of the promises made to him, however, he must stand responsible and bear the consequences. There were many suicides on Black Friday in 1929 when the stock market crashed.

Individuals who are Lithium sensitive can appreciate the exquisite internal feelings of another person's vibrations and can experience rejection without it being verbalized. The most intelligent are the most intuitive, and the ones who are most subject to depression and/or self-rejection. It is well known that amongst professionals, psychiatrists usually have the highest percentage of suicide. Lately, dentists have this dubious honor. Professionals, like most people, subconsciously absorb the vibrations of the people with whom they deal. Since depression is often a way of life today, and since these professionals deal very intensely with their patients, their exposure is of sufficient magnitude to have a devastating effect upon them.

Lithium sensitive individuals need people and human vibrations around them, but they may be shy. They may become retiring and distant because they are unwilling to make the first move when meeting others because they fear rejection. This shyness amongst people who need people further complicates the life of Lithium sensitive individuals. Their need for people becomes short-circuited by the fear of rejection. Although they may become shy, retiring, and withdrawn, this position in itself is equally uncomfortable. It represents another of the many paradoxes in which Lithium deficient individuals find themselves embroiled.

Of all the double binds which exist for individuals with circular mood disorders, the paradox between the fear of rejection and the fear of loneliness is one of the most exquisitely painful.

Antisappointment

It is rare that one has the chance to coin a new word. In a therapy session with a patient, discussing the anticipation-disappointment cycle, I said the word "antisappointment" rather easily, a signal that my Freudian slip was showing.

The anticipation-disappointment cycle is one of the most prominent features of a Lithium sensitive individual's orientation to life and causes them to experience many hurts on an almost daily basis. The majority of these are a function of self-inflicted emotional pain.

> *"What do you mean you didn't know I wanted another cup of coffee? If you really loved me, you would have known."*
>
> *"I know I said I didn't want you to get a cheeseburger for me when you were out. But if you really cared, you would have known I wanted one anyway."*

Statements such as these are commonplace. The lack of open and honest communication is one of the basic reasons families and relationships are destroyed. Only when individuals express their needs openly and honestly can they judge whether or not the individual with whom they are communicating is willing or able to meet those needs. It is difficult enough to deal with individuals when you know what's going through their minds; it is tenfold more complex to deal with them when you do not.

Earlier, we referred to the golden rule as we used it in psychotherapeutic practice. It is the expectation of another's behavior that creates an atmosphere in which disappointment runs rampant. To help prevent this from happening, each of us must explain to others what our needs are and why we are the way we are, otherwise we cannot expect them to tell us what their needs and thoughts are.

You can explain your needs to another person by demonstrating. If you need to be hugged, it is so much easier to walk over to the other individual, hug him, and let him know how good it feels, than to sit around feeling disappointment, waiting to be hugged, and feeling unloved.

There's a popular poster being sold now which usually has a tearful-eyed dog on it. The caption is, "God blesses those who expect nothing, for they shall not be disappointed."

Being a Perfectionist vs. Being a Pseudo-Perfectionist

The obsessive-compulsive individual is often thought of as a perfectionist. This is a supposedly brain type or "thinking" computer type, the polar opposite of a "feeler" type Lithium deficient individual.

Although individuals who are Lithium deficient are basically impulsive and oriented to the here and now, they may outwardly appear as perfectionists. Their need to be perfect is related to the fact that any less than perfect interpretation of a job they have done becomes a global attack on them as individuals. In order to protect themselves from feeling attacked and rejected, they must do a perfect job so they are not put down.

Any negative comment about the quality of a job performed

by individuals suffering from circular mood disorders becomes a negative statement about them as individuals. They become visibly uncomfortable and develop a set of feelings with which they are unwilling to deal. They perceive the attack as global and they feel put down, worthless, and hopeless, almost destroyed. Obsessive and/or compulsive behavior, although incongruous with their basic personality, becomes compatible with their immediate needs. They are not pro-perfectionism; they are anti–criticism. Individuals with a "feeler" personality must avoid criticism at all costs. They expend much time and energy going out of their way to avoid criticism, perfectionism merely being a byproduct. The term pseudo-perfectionism applies, rather than perfectionism itself.

When you disagree with individuals who suffer from a circular mood disorder, this disagreement becomes expansive and complete. If you disagree with their thoughts, ideas, or opinions, it is as if you disagree with them as people. The logic that follows then is illogical:

> *Not only do you disagree with what I say or do, but you disagree with me as a person.*
> *Therefore, you don't like me.*
> *Therefore, you have probably never liked me.*
> *Not only have you never liked me, but you are probably getting others not to like me either.*
> *You probably will never like me or leave me alone.*
> *You would probably like to see me ruined.*

This can become so extreme as to be clinically classified as paranoia. These individuals may then become either very withdrawn and retiring, or very angry and assaultive. The assault could either be verbal or physical. Individuals who become certain that they are being plotted against develop paranoia. They feel as if people are definitely against them. This complex situation can even evolve when you disagree with a Lithium deficient individual's opinion, or when you are asked if you like something they own, especially clothing.

To tell Lithium deficient individuals you do not like something they are wearing is tantamount to telling them you do not like them as people. That becomes a global and personal challenge, not just a discussion of clothing, but a personal insult. This is, of course, a very extreme reaction, but it exists and cannot be ignored.

Individuals who are subject to Lithium deficiency greatly need peer acceptance. They may wear clothing that is "in style," the style being that of their peer group. Since they are subject to peer pressure, they frequently and almost continuously wear clothing which is accepted by their personal peer group. The situation becomes more complex when you remember that Lithium deficient individuals have difficulty making minor decisions. I have seen in my practice that many adolescents and young adults have an array of jeans in their closet. By wearing jeans and a clean shirt, they do not have to make a decision and they can dress quickly. Since their peers think the same way, they also primarily wear jeans. There is a commonality to both their needs: not to make decisions and to gain peer acceptance by wearing jeans.

Fear of Criticism

It can be difficult to explain certain personality characteristics according to currently accepted psychiatric theory. It is obvious that each individual is unique, and his or her symptoms remain unique. However, one often finds a universal common denominator. The search does not have to be intensive since the number of common denominators is limited. It has been my overall clinical impression that the fear of rejection seems to be the most common factor.

Although I stress the "here and now" and problem solving approach to psychotherapy, one cannot totally discount the subconscious and preconscious states. Past experiences have an impact on the present that cannot be discounted. Whether they are easily retrievable as memories or are well hidden and defended against as part of the subconscious, we must recognize their existence. However, commonality of symptomatology and "here and now" therapy is still the most efficient and rewarding.

David's personality disintegration was overwhelmingly excessive and inappropriate to the circumstances and situations in which he found himself.

David was promoted at work. His income was in the upper middle class range. His wife had her own career and was equally, if not more, successful. Their combined income placed them in the top fifteen percent of the national average. They had no children and no responsibilities other than those they set for themselves. Despite this, they went very few places, lived modestly, and had an essentially Spartan lifestyle.

When David came into therapy, he was absolutely overwhelmed, suicidal, and essentially unable to function. He could not spend a day or even part of a day by himself. He was genuinely suicidal. His entire life was a shambles. He could not work, function as a husband or even care for his own most basic needs.

This was all essentially out of keeping with the present or historical data we could assemble. We took histories from him, his spouse, his parents, and other people. We established a very long and unusually arduous, painstaking, longitudinal history. Despite this, David felt no relief. He was placed on appropriate medication which did offer some relief, but his clinical course continued to wax and wane. He continued to have periods when he did unusually well and functioned almost normally, followed by other periods when he almost totally came apart at the seams. He frequently thought of suicide, and of existing rather than living, which was his self–imposed exile.

After much investigation, it finally became apparent that David "fell apart now so that he could avoid the rush of dealing with criticism later." David was so overwhelmed by the fear of criticism that his illness had to be dramatic. It could have been either physical or emotional, however it was only through the development of severe symptoms that people would not criticize him, but would rather pity and console him. This sounds like a rather simplistic analysis of a very complex case, yet it has held up under close investigation.

David literally extended his symptomatology to the ultimate (consciously or most likely subconsciously) just short of suicide so that no one could really criticize him without feeling guilty. He was so overwhelmingly concerned about criticism that his subconscious mind extended his illness to the point where he was emotionally crippled. This is a simplistic statement, but it is nonetheless true.

Another important point is that David, a college graduate, had certain ingrained concepts of psychiatry and psychotherapy that made therapy more arduous. Because he was intelligent, he understood that therapy was a time-consuming, ongoing procedure in which one must delve into the past to understand present difficulties. Because this was well established. I could not quickly or incisively destroy the myth. I had to work with him to establish full confidence and trust, so I could become much more incisive and help him. The longevity of treatment is usually proportional to an individual's socioeconomic and educational level. The expectations that they have been programmed into believing further complicate and extend therapy. To get well quickly would have been counter to David's basic need to suffer extensively.

David's suffering was of such magnitude that he lost much time from work and significant amounts of money as well. Before he got married, he had established himself as a "superman" at work. He did his own work and other people's, too. He took work home. He lived and breathed his job; no assignment was too difficult or complicated.

Eventually he reached his saturation point. He had no free time at all, no time left to live or enjoy the fruits of his labor. He could not even continue to complete the assignments he accepted. Mostly, he could not deal with the possibility of being criticized for not completing the assignments. His answer: a nervous breakdown—the only way to save face and avoid criticism.

David's circuitry became overloaded; his computer was overwhelmed. David's fear of criticism and, hence, rejection was so intense that self–rejection (suicide) was an acceptable alternative to external rejection. This self-created perfectionist's untiring image was about to be destroyed. Self-destruction was an acceptable alternative.

Fortunately, although his computer became disorganized, it did not become totally disorganized. His subconscious survival instincts prevailed. He became severely disturbed. His illness, though severe, remained treatable. Despite a rather long and shaky clinical course, David improved. He eventually returned to work and found that no one expected him to be a "superman." He wasn't criticized, nor did he continue to attempt to be a perfectionist.

David is continuing in outpatient psychotherapy. His verbal therapy is supportive and directive. The mainstay of his therapy is Lithium Carbonate.

David's case emphasizes that the response to therapy is not always dramatic. It may instead be slow and steady. Without Lithium Carbonate and its effectiveness in mood stabilization, all the verbal therapy in the world could not have helped David. Verbal therapy is viewed as an adjunct to Lithium replacement. Verbal therapy is important but not to the extent to which it has been publicized.

To Do or Not to Do

Individuals who are Lithium deficient are extraordinarily subject to peer pressure. They may become extremely decisive about major issues, can rapidly spend a great deal of money, buy a house on a whim, or a new car without a moment's hesitation. On minor decisions, however, they can become paralyzed. Generally, they are very indecisive and may spend a great deal of time figuring out which tie or blouse to wear. They are more than happy to lend someone large amounts of money on impulse, yet they become literally overwhelmed when they are asked to choose a restaurant for lunch or dinner. If the restaurant did not please the person sharing the meal,

Lithium deficient individuals feel literally destroyed. If someone else chooses the restaurant and the meal is not good, they will likely not even comment.

In the movie, *Marty*, when Ernest Borgnine, who played Marty, was talking to his friend, Angie, they were trying to decide what to do on that particular evening.

> *"What do you want to do tonight, Angie?"*
> *"I don't know, Marty, what do you want to do?"*
> *"Whatever you want to do, Angie."*
> *"Well it's okay, Marty, whatever you want to do."*
> *And so on.*

Individuals who are highly successful and intelligent may find themselves as little lambs meekly following the crowd.

When individuals who are Lithium deficient are placed in a leadership role, either by election or by circumstances, the situation is further complicated. Often without being aware of it, by subconsciously picking up other people's "vibrations" they try to make a decision based on their impression of the needs or desires of the people around them. They literally attempt to plug in their impressions of these needs or desires and are literally following their wishes as their leader. Essentially, they are playing follow the followers.

Indecisiveness

Lithium deficient individuals are most indecisive on minor issues. However as with all hard and fast rules, there are exceptions. Lithium deficient individuals either consciously or subconsciously set up other individuals in their environment to make decisions for them. Lithium sensitive individuals say, "Whatever you want to do is fine." Actually, this is not what they mean. They really mean that they want the other individual to make a decision which would make the Lithium sensitive individual most comfortable.

If a decision is made that is not within the framework of the Lithium sensitive individual, this is interpreted as rejection. Once the rejection button is pressed, the Lithium sensitive individual's computer reads "overload." Their reactions from that moment on tend to be illogical, irrational, and usually out of keeping with the situation at hand.

Since individuals who suffer from a circular mood disorder are unwilling and often unable to deal with rejection, they likewise assume other individuals have the same difficulty. Therefore they are reticent to initiate rejection. They tend to set up the individual with whom they are interacting to be the rejector. This appears to be a curious, yet ever present paradox. Although Lithium sensitive individuals are overwhelmed by rejection, their reaction to rejection is significantly lessened when they have either consciously or subconsciously set themselves up to be the one who is rejected. Lithium sensitive individuals have created a situation in which they have obviated their own responsibility to have to deal with the consequences of their own behavior. They have another individual make this most important type of decision, so they can then rationalize their own innocence in the decision making process.

Tom wanted to separate from his wife, Lois. He made her life absolutely miserable. He was verbally abusive, and criticized everything she said or did—the way she dressed, talked, or took care of the house or the children. After she was overwhelmed, put down, and attacked beyond endurance, she said, "Why don't you leave if you are so unhappy?" He replied, "If you want me to, I will." His reply was self-righteous and complied with her directions.

Picture for the moment two Lithium sensitive individuals who attempt either consciously or subconsciously to set each other up to make decisions. Most of these interactions tend to provoke anxiety and frustration. Neither individual is willing to make a final decision. Both of them are unwilling to live with the consequences of the decision making process. Relationships continue despite the fact that both people find it painful and even devastating.

These decision making concepts are further complicated by

the fact that Lithium sensitive individuals do not automatically plug into, or deal well with, the time frame of the future, so they naturally avoid situations that require making a decision.

Stroke—Don't Massage

Lithium sensitive individuals need to be stroked. Any situation or job is not completed until another caring individual comments or, hopefully, compliments them. They need the other individual to look at what they have accomplished and say, "That's a nice job," or, "You did really well."

If these compliments become too numerous, Lithium sensitive individuals develop a subtle sense of embarrassment. They walk a thin line between being stroked or appreciated and being embarrassed. They need their ego stroked but not massaged; they need to be appreciated but not catered to; they need to be cared for but not talked about; and they need to be held and to be allowed to hold. They do not wish to be made an example, only to be quietly appreciated.

When individuals who are Lithium deficient are offered a compliment, they frequently apologize for it. A female patient who recently visited my office was wearing an unusually attractive blouse and skirt. When she asked me how she looked, I said her blouse and skirt were unusually well-coordinated and up-to-date. She quickly countered by telling me she bought the blouse on sale, and she already had the skirt. She had a great deal of difficulty dealing with the compliment. She felt she was unworthy of such a positive statement, even though she set up the interaction.

Lithium sensitive individuals usually apologize for their appearance and/or the quality of their apparel. If the quality of their apparel and /or their basic appearance is not complimented however, they become distraught, anxious, and fearful. Lithium sensitive individuals frequently, if not continuously, need to be stroked, but not massaged. They need to be told that what they are wearing, or what they have produced, is acceptable, yet any overstatement becomes painful. Lithium sensitive individuals need to know you understand that their apparel and/or effects are socially and personally acceptable, and then the subject must quickly be changed.

Anaclitic Depression—Marasmus

The need to be touched and held enhances life. Many years ago, Dr. Rene Spitz demonstrated in a rather extensive study that infants who were cared for physically, fed and changed, but not taken care of emotionally, withered and died. Children who are reared in an environment which does not meet their emotional needs to be cuddled, held, and cared for in a loving, giving, sharing, manner wither and die. Children who were given minimal amounts of input often developed a "failure to thrive" syndrome.

A child who is afflicted with this syndrome is sickly and does not gain weight or grow according to the norms for his age. Lithium sensitive children require touching and being touched. They grow well under these circumstances, and an inability to have these needs met is overwhelming to their systems.

In order to further prove the universal application of this concept, Dr. Spitz and others did many experiments with animals. Newborn monkeys were placed in a cage with a substitute mother. There were several types of surrogate mothers. The first was a wire dummy similar to a form used for dressmaking. A bottle was placed in the breast area, and the infant monkey was fed. Many of these monkeys failed to thrive, and even died. Those who lived became angry and hostile, aggressive, and/or shy, withdrawn, and cowering. Other infant monkeys were placed in a cage with the same wire dummy, but a cloth was placed over the wire frame. These monkeys did better, but the best results were accomplished when the monkeys were reared by their mother.

Physical contact from a loving, caring parent is necessary

for the infant and for the parents as well. It makes them both feel good and they both thrive.

Children whose parents are emotionally sensitive also need to get and give hugs and kisses and frequently want to spend a great deal of time in another individual's presence. Our children do not want to be left alone, and they interpret being left alone as rejection.

You should readily identify this child's needs by recognizing them as a basic extension of your own. An individual born with these tactile, sensitive, touching, holding, feeling qualities should have these needs met on a regular basis.

When a child who is Lithium sensitive goes into a store, the parent might say, "Don't touch anything, Johnny." Because he is a loving, giving, caring person, Johnny quickly agrees, but the moment he enters the store, his hand reaches out almost reflexively to hold, touch, and pick up. This situation creates a great deal of difficulty for the child and the parents. The parent feels as if the child is being disrespectful and disobedient and is often hurt and feels rejected. The child is then punished for willfully disregarding the parents' wishes.

A Lithium sensitive child learns from tactile examination of his environment. The child is reprimanded for his poor behavior and uncaring attitude and is asked never to touch anything in that store again. This caring, pleasant child agrees immediately, and just as quickly, he forgets. He even forgot that he was just reprimanded for touching. A Lithium sensitive child, like a Lithium sensitive adult, is oriented to the here and now, to the moment. The reprimand he got just a few moments earlier no longer has any significant input.

Parents can easily avoid much of this difficulty if they recognize this characteristic in their child and feed rather than fight it. A child's tactile needs are ingrained within his psychological makeup, so it is easier to give the child something to hold and/or play with while in the store. A girl can carry a pocketbook with several pennies and perhaps a handkerchief in it. She can become creative within her own visual imagery and at the same time have something of various consistencies to touch.

A boy can use a ball, for instance, to hold and turn around in his hands to keep him active.

This is a general rule, meant to be followed and also meant to be broken. There are times when parents can reasonably expect their offspring to remain totally silent and not fidget. These time frames should be kept to a minimum

Debating Group

Lithium sensitive individuals have an intuitive knowledge of many fields. They can talk readily about sports, travel and foreign events, as well as mundane topics. They can readily, rapidly, and easily plug into conversations. Often, their knowledge is quite superficial, but they come across as quite studious. Lithium sensitive individuals are verbal and gregarious, and they are able to extemporaneously involve themselves in a multitude of areas. They have difficulty, though, planning a lecture and/or discussion. They do not usually do well in areas if they have to plan an outline of discussion or lecture. Spontaneity is their trump card.

Lithium sensitive individuals are quite comfortable dealing with unplanned events. They are relatively quick on their feet and can use "one liners" to engage in intense interactions with other people. They are reasonably quick-witted and spontaneous. If they become involved in a debate requiring logical inductive or deductive reasoning to substantiate points that are made, they become rather uncomfortable and even dismayed. They usually speak quite well and quite fluently. They relate easily to the needs of the group to whom they are speaking. They rarely, if ever, give the impression they are talking down to a particular group of people. They present themselves as being comfortable in all types of groups.

Always Is Not Forever

Lithium sensitive individuals are basically oriented to the moment. They react and relate to the here and now. They rarely plug into the past and have only a general overview of the future. When Lithium sensitive individuals say their statements are for "always," they "always" mean it, however, a few moments later is a new "always." They may give a new set of answers or responses, which are out of step with the previous answers or responses. To Lithium sensitive individuals, always is not forever; always is for the moment a statement is made. All of us have to be held accountable for the consequences of our behavior and for our statements, but should you place Lithium sensitive individuals on a polygraph, their statements of "always" would be validated. This makes it very difficult to deal with Lithium sensitive individuals on an ongoing basis. To them, "always" is a temporary situation.

A Knight in King Arthur's Court

To Lithium sensitive individuals, absolute honesty is secondary only to the feelings of rejection and loneliness. These individuals have a moral and ethical code like the knights in King Arthur's Court. Their moral and ethical code is firm and unyielding. It is of such quality and specificity that should anybody challenge a friend, to them it is as if one has thrown down the gauntlet. They accept the challenge as a personal affront to their sense of morality. This sense of morality is so specific that to Lithium sensitive individuals, a conscious lie is so devastating that it insults their very being.

Complications ensue. Lithium sensitive or "feeling" individuals would never intentionally "lie." However when asked a question, the "feeler's" response is of such quality and quantity as is necessary to please the questioner since fear of rejection is an overriding concern. When Lithium sensitive individuals are questioned, they give a response that has within it the kernel of truth necessary to substantiate the existence of an intense moral, ethical code.

"Feeler" individuals can respond to the perceived needs of another individual. They unconsciously and intuitively can perceive how an individual would like the question to be answered. To prevent rejection, they answer according to the perceived needs of the questioner. The kernel of truth exists, but the story may be slanted and/or embellished and/or exaggerated.

When other people ask the same question, that causes complications. Since the "feeler's" time orientation is to the here and now, the only question being answered is the one being presented at the moment. They do not plug into or relate to previous answers. Although their response is once again "the absolute truth," this truth varies from the previous truth. As each questioner asks the same question of Lithium sensitive "feelers," the answer is presented as if it is a gift for the questioner. Although dissimilar answers are being given, each answer, unto itself for the moment, is being related as the absolute truth. If a listener is able to relate his answer to another person's the answers may be quite dissimilar. When a large number of listeners get together, they may decide that the "feeler" has been "lying."

When these "feeling" individuals are involved in discussing issues or debating, they often either play the "devil's advocate" or they may take a role in the discussion which would basically please the discussants. Because Lithium sensitive individuals are intuitive, imaginative, and verbal, they are able to easily take any role in a debate which feels right at the moment.

These individuals may literally be able to debate both sides of the question with equal ease and equal knowledge. This knowledge is usually hearsay and/or tailored to meet the needs of a group at hand. They may literally debate a specific issue in a multitude of directions with any number of groups with which they are involved. When these different groups and members of these groups get together to discuss the role taken by any particular individual during their discussions, Lithium sensitive individuals may outwardly appear to be "lying." This is to say they have overemphasized their verbal abilities and overexercised their intuitive abilities and once again put themselves in a situation which to all outside appearances places them in the role of "liar."

All thoughts, events, feelings, and situations that Lithium sensitive individuals eventually find themselves in are almost always counter to their beliefs, ideas, original plans, and desires. Often, what happens overshadows what they intend to say. It is often appropriate to the moment but contrary to the basic tenets of the Lithium sensitive individual.

"Feeling" individuals almost always mean well. They usually try to please everyone, and most often they please no one. They paint themselves into a corner and then must suffer the consequences.

The Popeye Syndrome

Lithium sensitive "feelers" require a human vibration of the opposite sex around them in order to feel whole, complete, and intact, to feel entirely comfortable, and to be able to reach their maximum potential.

For the cartoon character Popeye to be able to function at his maximum potential, to reach unusual heights of capabilities and functioning, he must eat his spinach. Without it, Popeye is like everyone else. With his spinach, he is able to break free of chains and other encumbrances. He is able to attack and subdue Bluto. He can once again win his love, Olive Oyl. As Popeye needs his spinach, so it is essential for "feelers" to have another human vibration within their environment in order to function. This vibration ideally must be of the opposite sex. A vibration of the same sex is better than none, but that is settling for second best.

The need for another human vibration of the opposite sex becomes so intense and involved that it creates problems for average "feelers." They regard their need as a weakness rather than as a trait or personality characteristic. If they view it correctly as genetic, they can take good advantage of it. Since it is most often viewed as a negative, it creates internal stress for "feelers," and intense, painful alterations in interpersonal relationships. It places the needed individual simultaneously in an advantageous and disadvantageous position.

Lithium deficient individuals need tactile input. People who need Lithium are visceral and tactile, the "teddy bears" of life. They are individuals who need to kiss and be kissed, to hold and be held. If their loved one were to bring home one flower, they would view that as an overwhelmingly positive event. They don't need a dozen roses, just one. They don't always need to be held and hugged, but should you touch them on the shoulder while they are at their work bench or washing dishes, this input is dramatic, intense, and exceedingly positive.

When a "Feeler" Falls in Love

Individuals who are Lithium deficient have a very rich fantasy life. In their own minds, they often create a set of circumstances, feelings, and thoughts which are so vivid and intense that the fantasy aspects often become reality to them. This fantasy life, these thoughts, ideas, and feelings become so entrenched, that they may lead to such an indelible message and impression that they are, for all intents and purposes, real.

Each of us has what is known as an "ego ideal." We can establish within our own framework and our own ideas a set of thoughts, beliefs, and concepts of what we want to have, hold, and endear in another individual, and these thoughts become enmeshed with what we "see." Each of us has a concept of the perfect spouse. When we meet an individual who, for whatever reason, initiates within us a certain feeling or vibration, we suddenly and inexplicably endow this individual with the qualities attributed to this "ego ideal," as if it really exists. It then becomes an ultimate reality in our minds. This positive initial chemical interaction, this vibration, this feeling, can incorporate all these aforementioned qualities. In an attempt to solidify this concept, when one "falls in love with another individual," we may for the moment refer to this feeling of "love" as H-E-A-T for being "in love." We are in an intense relationship, the intensity of which exists only within our thoughts, ideas, feelings, and our own minds. As we fall in love (H-E-A-T), we fall for another individual in such a manner as to imbue them with all of these magnificent, wonderful, intense, yet usually nonexistent, qualities. We see him or her the perfect mate, the perfect individual, and the individual who has all the qualities we desire and need, but as time goes on, we suddenly see that this individual is not quite as "caring," not quite as "thoughtful," and not quite as "giving" as they were. The fact is they were never as thoughtful, giving, or caring as we initially saw them to be; they were just themselves. Since they have always been just themselves, this is a very confusing situation.

Don Quixote in the Cervantes novel saw Dulcinea, a charwoman who worked in a halfway house as a kitchen helper, as his lady. He took the barber's shaving bowl and believed it to be his golden helmet. He saw Dulcinea, who was sometimes a "lady of the night," as his special lady and imbued her with the attributes of a royal lady and ascribed to his barber's bowl the qualities of a king's crown. Lithium deficient individuals place upon another individual romantic qualities and create an image of what they want to see, rather than what really exists.

This sense of wanting and needing is so intense and overwhelming that it makes a fantasy an intense, obvious "reality," This "reality," which never existed, still does not exist. With time, the intensity of the giving, caring, and loving fades away, but they never really existed at all.

Lithium sensitive individual are so giving, intense, loving, and caring that they see only what they "feel" they want to see. However, with repeated insults to the system, reality eventually sets in, with great emotional pain. This pain becomes so intense that the sufferer eventually withdraws. Once again, they become disillusioned with the fact that people "change."

They are unwilling or unable to recognize that these people have not changed, but rather they now see the reality.

Lithium deficient individuals set themselves up for great pain, anxiety, and depression. Their relatively simplistic view of the world allows them to see the world as they want it to be, as they wish it to be, and, therefore, as it must be. However, reality is not as they want or wish it to be. Cervantes' "Dulcinea concept" is almost universal. This absurd situation comes under control only with the application of small amounts of the mineral Lithium. Since "feelers" are here and now oriented, when they meet a new individual they are likely to fall prey to the same set of circumstances. Others may say they are making the same mistake again, but they are not. Since they are oriented to the moment, they are *making a new mistake again*. These statements are ambiguous; however, if we think about it in detail, or feel it in detail, you can easily see that these incongruities readily coexist quite comfortably. They follow a comfortable pattern of illogic.

I Need Him (Her) So Much That I Hate Him (Her)

It should be obvious by now that there are many paradoxes with reference to the needs of Lithium sensitive "feeler" individuals. These people have a biological, hereditary, inborn need for another human vibration in their environment. This need is all-encompassing and all-meaningful. For many, it is as intense as a diabetic's need for insulin. Without another human vibration in their environment, the average Lithium deficient "feeler" will do almost anything to maintain a degree of interaction with another human being and prevent withering.

We can now view from another perspective those individuals who were regarded as weak and inferior because, after they are beaten, they return to the individual who has beaten them. I have seen females report to hospital emergency rooms with a broken bone. After the bone is placed in a cast, the female often returns to the person who beat her. I could not understand that for a long time, but when you weigh the emotional pain of being without another human vibration in the environment against the pain of physical injury, the pain of vibration deprivation is far greater. Often, an individual who is the victim of multiple beatings will still return to the source of those beatings rather than experience the ongoing pain of being alone.

This situation becomes more complicated. Individuals who need other people's vibration then become angry with the other individual because of this dependency situation. Lithium deficient individuals may become overwhelmingly resentful and even hateful toward the individual upon whom they depend. They blame that individual for the dependency situation rather than recognize it as their own personality characteristic.

People may think this information is double talk, but the truth is just the opposite. This information is so intense that each and every one of us should stop, think, and reevaluate our closest associations.

The situation becomes more complex, because there are several variations on the theme. The intensity of the dependency continues and may even increase with age. Lithium Carbonate is dissipated from the body as we get older. As "feeler" individuals continue to have their Lithium Carbonate stores decreased, their need for another human vibration increases and they overreact to even minor insults. Lithium deficient individuals become even more sensitized to the behavior of those around them. They become so intensely dependent on the vibrations of their mate that should the mate return home late from work, this is enough of a sense of rejection to create an argumentative and hostile atmosphere. It can become so intense and painful that the need creates hatred, animosity, disdain, disgust, and even outright warfare with the individual who is needed.

The mate in need who waits at home paces and becomes ill at ease and anxious. This individual becomes so overwhelmingly uncomfortable that he or she is literally unable to sit still, and feels excessively restless. When the mate does return home, even ten minutes late and perhaps even carrying a gift for the loved one, the individual at home attacks, questions, berates, and interacts negatively with the individual who is returning home.

What "feeling" individuals really want and need to say is how much they love and miss the mate who has just returned home. Instead of saying, "I really missed you today. I love you," the greeting is negative, hostile, painful, and at times, even frightening or physical.

This situation shows a positive can become a negative. A well-meaning, innocent situation can become hostile, aggressive, and painful. The study of a Lithium sensitive individual is like "Catch 22." That is when you are "damned if you do and damned if you don't." Lithium sensitive individuals suffer so intensely from such a multitude of situations that at times it appears there is almost no way out. Unfortunately, they tend to create unusually painful situations from which they have great difficulty extricating themselves.

The intensity of their need for another individual is so overwhelming and of such magnitude that they actually become angry and frustrated with the individual they need. They deny their need instead of acknowledging it. Through this denial, they become increasingly frustrated, hostile, and self-defeating.

The expression, "I need him/her so much that I actually hate him/her for it," applies perfectly to Lithium sensitive "feelers." Individuals who need other people's vibration actually begin to believe they are victimized by the other individual, and that it is the other one who makes them feel dependent. Reason, logic, and understanding are unimportant factors in the subconscious and/or preconscious understanding of a "feeler's" needs. We must always remember that the important factor is the overall feeling tone of a situation.

Lithium sensitive individuals enjoy their privacy. They occasionally like being alone, but only when it is their own choice. If they are placed in a situation where they are left alone and/or another individual does not arrive on time, anxiety, tension, frustration, an internal pacing, a racing feeling, and a sense as if they are crawling out of their skin develops. This anxiety is both self-destructive and destroys the relationship. The intensity of the interaction becomes negative rather than positive. Although they have difficulty acknowledging it, Lithium sensitive individuals will wither without another's vibration in their environment.

To "feelers," this other human vibration is absolutely necessary and essential for their support system. This is absolutely necessary for their survival, sense of well-being, and sense of being a whole and complete individual. This situation becomes crucial in therapy since one must recognize that manipulation or alteration of the external environment of Lithium sensitive "feelers" is a crucial and an essential part of their therapy. A simple, straightforward manipulation of the external environment of Lithium sensitive individuals is not only crucial to therapy, but that therapy in itself would not be successful without it.

Lithium sensitive individuals need another vibration to complete a task more comfortably. Walk along while this person mows the lawn and watch how fast the job gets done. Frequently comment on how well the individual is doing and watch the job get done correctly.

The external environment—a mate, parents, children, friends—is in fact more important than individual psychotherapy itself. "Feelers" regard their biological need for another vibration as a weakness and a negative influence. It is important to remember that this situation is but one of the many characteristics of their personality style. They must learn that this style is the most common of all personality clusters. I will reiterate and stress that at least seventy-five percent of the population is Lithium deficient and maintains this particular personality characteristic as one of their own. As with any other biological-physiological phenomenon, the degree to which one is dependent on an external source for feelings of wholeness follows Bell's curve of normal distribution. There is a fluctuation of need which parallels other fluctuations in human characteristics, such as body weight and height.

Lithium deficient "feelers" can be pictured as having the

capabilities of Superman, but Superman appears with Kryptonite in his pocket. He usually has an inordinate, unusual ability to deal with any and all of life's stresses. He can conquer the world and offer worldly answers to life's problems, but with Kryptonite in his pocket, he sits essentially motionless and incapable of even tying his own shoes. Lithium sensitive individuals are likewise capable of great strengths and great accomplishments. Without another vibration in the environment, "feelers" have their own form of Kryptonite in their pockets. With the addition of another human vibration to their environment, "feelers" can function effectively and consistently. They are better able to reach their maximum potential and abilities.

The ability to function and the capability to do so, therefore, varies significantly. Their ability and capability become more in tune with each other when the vibration of the other individual is present. Success only succeeds when those around them appreciate, understand, and can feel the necessity for another individual's presence. I regard both Popeye and Superman as clear-cut examples of the need for positive vibrations for success. Without spinach, Popeye could not succeed. Without the removal of Kryptonite, Superman could not, either. Without the presence of another caring person, "feelers" often, if not always, are unable to fulfill dreams, desires, or even a vocation.

How to Get a Difficult Job Done Quickly vs. an Easy Job Done Slowly

Lithium sensitive individuals have a biological need for another human vibration in their environment. Give them a six-hour project to do, allow them to have another human vibration in their environment, and they will complete that project in well less than half the time allotted. But if you request the completion of a minor fifteen-to-twenty minute task and leave them alone to do it, there is little chance it will be completed.

When we offer the concept of being alone, we refer to an involuntary or imposed time frame. Lithium sensitive "feelers" who elect to and find solace in being alone, can handle that time adequately and comfortably. But if they are left alone when it is not their choice, they have a great deal of difficulty dealing with this aloneness. Lithium sensitive "feelers" will pace, turn on the television to see another human being, make many phone calls to hear other voices, even long distance, unnecessary calls. They may turn the television on and then turn it off, open and close the refrigerator, and walk outside and check for the mail even though they know it is not there. They become uncomfortable, uneasy, and frustrated. The paradox then exists when their mate returns. Instead of saying, "Gee, it's good to see you; I missed you," they get angry, hostile, aggressive, and negative. The anxiety they feel builds up to an extreme and they "blow up."

This is very important to consider when we deal with Lithium sensitive children. Children who have one or both Lithium deficient parents often are unable to do simple things. Upon leaving the house a parent may instruct the child to "make your bed and clean up your room," but when the parent returns the work has not been done. The parent becomes hostile, angry, aggressive, and frustrated and treats the child as if the child is unwilling to respond to simple instructions and the requirements of the household. The opposite is quite true. Just as a Lithium sensitive parent is unable to complete a task when left alone, so too is a Lithium sensitive child. A Lithium sensitive child could help make eight or ten beds in succession without any argument, provided the task is done with one of the parents or siblings. If left alone to make just one bed, it is an awesome responsibility and task which is often, if not generally , not completed. The individual who does not complete this task then develops feelings of self-recrimination, inadequacy, inferiority, and worthlessness.

Instead of fighting the needs and requirements of Lithium sensitive individuals, if would be easier and more expedient to feed the requirements by offering to stay with them while they do their chores.

Rent a Friend or Rent a Vibration

I have treated many patients who have had significant difficulty interacting with their mates because they "did not do the housework." I have met women who literally have to use every possible amount of effort to make the beds and do the dishes. They sit around, watch television, and talk on the phone all day. These people are seeking another human vibration via the phone or by watching television, especially soap operas. The housework does not get done, and when their mate comes home, there is a sudden frenzy. They are rapidly and efficiently doing the housework and then rushing to prepare dinner. Arguments ensue when the working mate assumes the nonworking mate has intentionally wasted the day and is purposely doing housework when they could be spending time together and enjoying each other's company.

The working mate would ostensibly appear to be correct, but the truth is that the nonworking mate is so overwhelmed by being alone that the ability to perform routine daily functions ceases.

I must state once again, and emphasize that *these personality traits are innate, inborn, genetic characteristics* of Lithium sensitive individual, not to be fought or denied, but to be worked with, understood, and mastered. They are not to be regarded as weaknesses, but as a fact of life. They must be confronted and discussed between partners to mutual satisfaction.

The Self-Made Penniless Millionaire

Leonard is a forty-year-old married white male, father of four. He has three adult children by his wife and one young child by his estranged mistress. Leonard comes from a lower middle class background. He has always been a hard working, hard driving individual. He married his childhood sweatheart, a caring, loving, devoted, nondemanding individual whose lifestyle was totally geared around her husband and his needs. At a young age, Leonard became a millionaire. He devised a hugely successful system for inks and their application.

The chromosome for the genetic transmission of intelligence and creativity also appears to house the counterbalance: a chronic sense of worthlessness, inferiority, and insecurity. Leonard, who is outgoing and gregarious, would spend his money freely and in a relatively flashy way. Because of his background, he was basically family oriented and took excellent financial care of them, but many of their emotional needs were not met. As a typical Lithium sensitive individual, he set outlandish goals for himself and for his company which he was able to carry out. As he became more successful, he spent more lavishly and began creating an image of himself which he would carry out despite personal hardship and loss. He would then get himself involved in auxiliary businesses which were sometimes extremely successful and other times, horrible failures. Despite his success, Leonard remained gullible and naive, the prey of any outlandish scheme. He needed greater and greater success, which became his temporary downfall. He lost sight of his goals and became blinded by his own drive and caught up in his own treadmill.

He merged his company with another and because of various intricacies, he sustained great financial hardship and essentially went bankrupt, while the other officers of the company profited greatly.

Just before his downfall, Leonard became involved with another woman for the first time in his life. He fell in love with her despite his cognitive recognition of her multiple frailties. He was Don Quixote, and he met his Dulcinea. There was a chemistry between them which made him feel as if all his fantasies were being fulfilled. He recognized she was a prostitute, yet she was still, without question, his one and only lady.

For a time he lived a double life. Since he always got up and went to work early, his wife, a very naive, trusting individual whose trust blinded her to reality, was not suspicious. When she finally became aware of the situation, she confronted him. He left her and moved in with his mistress who became pregnant and had a daughter. The conflict that followed became even more intense and bizarre. Leonard's wife remained his closest ally. He contacted her infrequently and spoke to her rarely, but whenever he was ill or in need of emotional support, he would turn to her and she steadfastly remained by him.

With time, Dulcinea began to show her true colors. She once again became very hard, nasty, cruel, attacking, insulting, and demeaning. She would defame Leonard's masculinity, background, and earning power. She would attack him to the very substructure of his own being. Cognitively, he recognized that there was a change in the relationship. His once loving, giving, caring, sharing, loyal "lady" was no longer the same person. As one looks at the reality of the situation, it is clear the intensity of their relationship was in direct proportion to the size of his bankbook. When she first met Leonard, he was riding the crest of his own financial wave. Money flowed freely. As his income diminished, so did her "true love."

When Leonard and Dulcinea began their relationship, the hours he worked, the trips he took, the things he had to do were accepted and expected. As the relationship became more solid, she resented his business and the time he spent away from her. Despite the fact that his working was the foundation of his financial success, she refused to allow him to go to work. If he did, she would call him so frequently, so incessantly, and so intensely, that he had to return home. She did not hesitate to thoroughly, completely, and entirely embarrass him in front of his friends, at business meetings and in front of his employees, as well as in front of his own family. As she continued to erode his self-esteem, self-confidence, and desire to remain creative in his field, his income decreased in proportion and for a while was essentially nonexistent. She became even more defaming, aggressive, demanding, and demeaning.

There is now a very vicious cycle: the energy level that was his productivity quotient became his enemy. The amount of energy he spent working was unacceptable to the person who wanted as much money as she could get from him. The only way Leonard made money was through his own efforts, so we see the creation of yet another double bind. Our "hero" was forever in a position where he was "damned if I do, and damned if I don't." There are so many examples of how one's success can be absolutely destroyed by the very same individual for whom the success was meant.

The crucial point to remember is that on the same chromosome with the gene for intelligence, creativity, and imagination appears to be the counterpart, the gene for inferiority, insecurity, and worthlessness. The more successful individuals are, the more worthless they can feel they are.

Leonard's story is a classic example of how crucial it is for Lithium sensitive individuals to surround themselves with caring, loving, and giving people. Lithium sensitive individuals, regardless of the heights of success, have a trememdous biological need for external support. Their locus of strength and power comes from an external support system. When Lithium sensitive individuals have negative, external support systems, hopelessness, and worthlessness dramatically increase.

Despite the negative, hostile, and tense situation which developed, Leonard was still unwilling and/or emotionally unable to disengage from the life he chose. For years, he continued to see only good in his "lady." He made excuse after excuse for her negativism, hostility, and activities. He waited on her hand and foot, for which he received even more disdain and disgust. As the cycle of his mood swings started bringing him back toward an improved mood, he was able to disengage from her on a full-time basis. He returned home and received the nurturing and caring he sought, which enabled him to once again begin a whole new, successful company. He still has not been able to fully cut the ties between himself and his Dulcinea, though.

Intelligence, a high energy level, an ability to succeed and to create are not unto themselves a set of all-powerful personality traits. They must be tempered and worked with, and above all one must carefully select the external support system that one is willing to relate to and deal with so that a situation such as our "hero" and his "Dulcinea" can be prevented. Once it begins, its intensity, longevity, and effect on one's person are unpredictable.

The Reverse Dulcinea Syndrome

The Dulcinea syndrome related to the fact that individuals have created within their own feelings a picture and concept of the perfect mate, spouse, or friend. They are then able, without hesitation or thought, to place upon another individual all of the concepts, feelings, or ideas that they have created. Conversely, each of us has also established within our own framework of thoughts, ideals, feelings, and needs a concept of ourselves that we would like to project to others, so that they will perceive us in the manner we will find most comfortable.

Consequently, we establish an image of ourselves as we would like, and almost require, others to perceive us. Once established, this image must be maintained at all costs. When we refer to the penniless, self-made millionaires, we see individuals who have achieved personal and financial success. During successful periods, they are more than willing and able to pick up the check at dinner and to supply ideas, money, and advantages to their court of jesters (their followers). As the fortunes of business change, so does the income of self-made, penniless millionaires. Since they have already established themselves in a role, they must support it. Despite financial misfortune, they may use their very last penny to maintain the self-created situation. Despite the fact that the individuals and their families may suffer, and that the chances of rebuilding the business may diminish, they will maintain this image at all costs and despite personal losses.

A Lithium sensitive individual is one whose rational thoughts and ideas can easily and readily be overwhelmed by feelings and by self-created and self-inflicted expectations.

Once an image is created, it must be continued. The creation is self-enhancing at its inception, but at its conclusion it may well be self-defeating and destructive. Lithium sensitive individuals who have created an image of holding onto the chain of an anchor, despite the fact that it may have been disengaged from the ship, could in all sincerity hold onto the chain as the anchor pulls them to their own demise.

The Cassandra Syndrome

According to Greek mythology, Cassandra was the daughter of Priam and sister of Paris who abducted Helen of Troy and created a situation which gave rise to the Trojan Wars.

Cassandra created an interesting paradox. She pleased many of the ancient gods but unfortunately, she displeased almost as many. The gods she pleased were so pleased by her that they decided to give her a gift. They bestowed upon her the gift of absolute prophesy. They told her there never was, is, or would be another mortal more expert at prophesy than she. Once a gift is given to a mortal, the others gods cannot interfere.

The gods she angered had to accept this as her destiny, but they bestowed upon her a "gift" of their own. Although she would be the most expert prophetess that ever was, is, or would be, they stated and cast upon her the fate that nobody would ever believe her.

This situation could cause a mortal overwhelming pain, turmoil, and anxiety.

Average Lithium sensitive, emotionally labile overreactors have been granted the gift from God to have above average intelligence and the ability to be creative. Conversely, they have also been given the curse of an inborn, ever present but varying sense of inferiority, insecurity, worthlessness, and helplessness. The individuals who are essentially the most intelligent are essentially the most insecure and gullible. These feelings of helplessness, hopelessness, and insecurity represent an overriding haze or cloud of gloom over the personality of the outwardly successful.

It is like an exquisitely balanced seesaw: one end represents intelligence, achievement, imagination, and creativity. The other side, in equal balance, represents inferiority, insecurity, helplessness, and hopelessness. It is well known that the most gullible and easily conned individuals are doctors, lawyers, and dentists since they have a dramatic need to be liked by and to please even the con man.

Intelligence, caring, loving, giving, and financial success parallel gullibility and naiveté. The most successful are often the most self-destructive. They give to a fault and their giving often represents buying another individual's love and affection. The more they give, they more they expect in return. They literally force the person they give to into a position where they must "stroke" the giver. Their need for strokes and reinforcement is not a desire, but rather an overall genetic, inborn, biological need. Givers give to fulfill their own needs and not those of the receiver. They attempt to overcome their insecurities by demonstrating worthiness by providing "things" for the receiver. If, per chance, the receiver is not grateful for the "gifts" bestowed upon him, givers develop hurt, angry, often hostile feelings. Already insecure, they become outraged. This intensity of mood usually passes quickly and peace is restored. The receiver usually apologizes for not being properly grateful and the apology is accepted.

Another paradox becomes apparent. Givers are usually more forgiving of acquaintances and business associates than of family and close friends. Loved ones want to please, but because of the frequency of mood changes and the ensuing tension, the givers can never be pleased consistently. Additionally, they recognize that their loved ones are aware of their instability and remain fearful of this being used against them.

Intuition—Gift or Curse

Individuals who are gifted with intelligence, creative visualization, and intuition are simultaneously cursed with the lack of trust in their own intuitive powers. Most individuals who are gifted with intuition generally refuse to follow their intuitive directives. If they were to review their own decision-making processes, they would find that if they followed their initial intuition, most of the time they would have made the correct decision.

When Lithium sensitive individuals attempt to intellectually or cognitively override their intuitive powers, their course of action is almost universally doomed to failure. Why Lithium deficient individuals refuse to follow their intuitive direction is beyond logical description. We must always remember that logic rarely wins out over intuition. Suffice it to say that Lithium sensitive individuals would be significantly more satisfied in the outcome of their decision-making processes if they would allow themselves to follow their intuition or "gut reaction."

Parapsychology

Along with other characteristics of individuals who are considered "feelers" or Lithium sensitive, there is often a more than usual interest in parapsychology. Yet not each and every individual who is a "feeler" is one who has an intense interest in parapsychology.

It is an awesome task to think of the brain. There are billions of cells connected to each other with even more billions of connecting terminals. To simplify, see the brain as a complex computer with multiple transistors and circuit boards. These transistors and circuit boards are interconnected and at different times have varying degrees of prominence. These transmitters of information are aligned to the network of transmitting substances which send the messages from one brain cell or group of cells to another. There is a "sodium-potassium pump," an exchange of sodium and potassium from without to within the cell that establishes a mini-electrical discharge which is then sent to other cells as a message. If we view this mini-electrical discharge as the transmitter, we can understand that another individual's "receiver" may pick up this message. Gifted individuals are those who have a multiple channel, perhaps 400–1,000 channel receiver, akin to a simple CB sender-receiver. So telepathy is nothing more than a gift. It is the ability for one's receiver to pick up a mini-electrical discharge sent by another's transmitter. Clairvoyance, the ability to see what is happening elsewhere, perhaps even before it is being seen, along with precognition (knowing something is going to happen before it does), are other examples of transmitter-receiver phenomena. Clair-audience, the ability to hear what is going on elsewhere, follows the same parallel. People who are born with the Lithium sensitive type of physiology are those who are most "talented" in this area.

A more practical and frequently "felt" circumstance is intuition. Lithium sensitive people feel; obsessives think. Lithium sensitive people have been both gifted and cursed with the ability to receive others' vibrations almost at will. Lithium sensitive "feelers" can walk into a room feeling good; yet, should the individuals in the room be despondent, the "feelers" will rapidly absorb these feelings and become depressed, almost like a deflated balloon. The opposite is also true. Should Lithium sensitive "feelers" feel down and enter a room where people are very up, they may also become "up." Unfortunately, it is easier for "feelers" to respond to a depressive than to a positive environment. Individuals who are down may come up, but at a much slower pace and a much lower rate.

Lithium sensitive "feelers" may be asked a question about a certain statement they have made. It is most difficult for them to logically explain the "feeling" they have. Their thought processes are unusually rapid and conclusive. Their answers are absolute and yet do not follow any logical sequence of problem solving techniques. They often have a great deal of difficulty putting into words the feelings they experience. This inability can make them appear somewhat dull and less than objective in their reasoning."Feelers" are individuals who get in the most trouble when they attempt to cognitively or cerebrally override an intuitive feeling. Because of their romantic nature and desire to perceive things as they want them to be rather than as they really are, many "feelers" frequently say, "I don't feel the situation is right, but I will try to correct it." Despite the intensity of their desires and their own good intentions, they fail to change the situation to the way they feel it should be. It is this constant attempt to override their intuitive feelings that creates pain and discomfort for most "feelers." If, instead, they would view a situation as it really is, as they feel it is, rather than as they would like it to be, many of their problems would disappear.

The introduction of a psychic, an aura reader, or an expert in the tarot cards, brings excitement to most "feelers." Despite their level of educational achievement and/or professional ability, many have a separate, distinct, and more than casual interest. This may seem incongruous to a casual observer, but it is understandable because it "feels" right to the "feeler." Many "feeling" individuals actually alter their personal and/or professional lives by incorporating parapsychological information. There is a certain sense of comfort in knowing the unknown, a certain sense of excitement and simultaneous anxiety when dealing with the occult. Since occult means existing but barely perceptible, it fits in with "feelers'" personality constellations, where their feelings are perceived but most often are not understood. They perceive before they understand. We all have the genetic birthright of developing parapsychological abilities; the degree to which we develop them is up to us.

Thought Abbreviations

Lithium sensitive individuals talk in abbreviations. They offer segments of sentences, thoughts, and ideas, and assume that the individual with whom they are relating can fill in the blanks. This concept creates frequent ambiguities and misunderstandings. The abbreviation or condensation of Lithium sensitive people while they speak does not necessarily relate to the receiving system of the individual being spoken to.

They are not only speaking in abbreviations, but they are also receiving and responding in abbreviations. There is a great deal of conflict within this system. Lithium sensitive individuals usually respond to a question or statement, but the response is often less than adequate. They respond approximately. When I asked, "Were you in the service?" one patient replied, "Yes, I was discharged." When another individual was asked, "How far is it from here to Boston?" the response was, "Yes, the car is filled with gas." The response of Lithium sensitive individuals is less than accurate to the question asked.

Lithium deficient individuals respond to the part of a question that interests them most. They often lose sight of the overall question at hand. Their focal attention is so acute that they respond to a segment of a statement without "hearing" the rest of what is being said. This spontaneous trance or daydream state that most Lithium deficient people frequently find themselves in causes a great deal of friction and many arguments.

"I told you that before!" Donna says, absolutely sure she is right.

"No, you most certainly did not!" says John, with equal certainty. A verbal battle ensues. Both Donna and John would in all likelihood pass a lie detector test, but neither one of them can accept the other's position. Each being Lithium deficient, they both now have feelings of being mistrusted, disbelieved, and, therefore, rejected. Hurt feelings develop quickly and tempers flare. The noise level of the discussion rises and nothing is accomplished. Eventually both individuals realize that a forced compromise or apology is necessary. The individual with the greater level of sensitivity (or fear of rejection) usually apologizes first. Both continue to remain angry but are afraid to show this anger directly.

Their inability to be assertive or open and honest in a conversation markedly constricts options for self-expression. A more positive way of handling the example above might have been by saying, "It doesn't matter whether or not you remember telling me. I don't remember hearing it." Or, "You say I didn't tell you. I seem to remember I did. Nonetheless, I'll say it again."

People would not be people if they spoke this way all the time. However, it would be a refreshing change from having frequent, intense, and predictable arguments.

Logarrhea

Often, Lithium deficient individuals greet you with "How are you?" and then proceed to tell you how they are. What they are really saying is, "Hi, you are fine. How am I? Well, thanks for asking. I'll tell you."

When Lithium deficient individuals are on a high or relatively high mood, they are very, very verbal. They may ask a question and then cut you off in the middle of the answer and proceed to answer the same question as if you had asked it of them. Lithium sensitive individuals do not intend to be rude or ill-mannered. Their need to speak is merely pressing another button on their computer, which allows them to continue to be very, very verbal.

Lithium sensitive individuals often suffer from Logarrhea, diarrhea of the mouth. They talk excessively and often with a sense of pressure to their speech pattern. They give expansive answers to questions. They are very verbal and yet rarely answer the specific question which was asked. They often get anxious when you try to pin them down, and they become angry when they are asked to be brief. They frequently interpret requests to be concise as rejection, and we all know what happens when the rejection button is pushed.

Binge Behavior

Binge spending is a very significant type of "feeler" behavior. Lithium deficient individuals are capable of spending overwhelming amounts of money in a short period of time. Their spending may cripple the family for extensive periods of time. This kind of spending applies to all socioeconomic groups, but is usually geared according to the level to which Lithium deficient individuals are accustomed. The woman who buys herself a new pair of shoes and a hat when her children don't have enough to eat is just as much a binge spender as the woman who spends thousands of dollars on clothes she can afford but probably doesn't need.

Lithium deficient individuals do not spend because they need what they buy. One reason they spend is because they need to "feel better." They "feel" as though spending will elevate their mood if they are "down," or satisfy their urge to do something expansive if they are "up." No matter what the reason, the result is often tragic when these items have to be paid for.

Many Lithium deficient individuals become hobbyists. They get deeply involved in any new hobby that catches their interest and after having a relatively good idea about the nature and quality of this hobby or sport, they purchase each and every piece of equipment possible. They usually purchase the finest, most modern, and essentially the most expensive equipment. However, even before the last piece of equipment has arrived, they may rapidly lose interest in the hobby or sport and instead embrace an entirely new hobby or sport and buy its equipment, claiming this time they "will stick to it."

Lithium sensitive individuals are basically likeable when they promise their mate that this is absolutely the last binge-type situation, the mate usually believes it. Even though this present hobby may seem to repeat a previous pattern, he swears it is not. "This is the only one I ever really wanted," he says, however, the pattern continues.

The social attitude of Lithium sensitive individuals is likewise a type of binge behavior. They may stay with one or another of their friends for an extended period of time and suddenly leave them for another friend. They may be into eating out once or twice a week for months and suddenly stop eating out. They may suddenly decide to see a Broadway show and then wind up seeing every show on Broadway and then not return for years.

Lithium deficient individuals are extremely impulsive, impetuous, and are locked into the moment. They are so subject to peer pressure that they could actually be involved in a lynch mob and the next day not realize it or even be able to explain how it happened. They are caught up in the moment and with the vibrations of the group with which they find themselves.

They absorb, they share, and then they become part of it. L
ium sensitive individuals can readily absorb the vibratio
the people around them. They can become depressed when the group around them is depressed; they can become excited when the group around them is excited. They can be so absorbed in the feeling of the moment as to literally be involved in murder or gang rape without being actually able to explain how they became involved. The explanation is that their subconscious mind allowed them unwittingly to share in group process or group hysteria.

"Auction fever" is an example of group hysteria. Lithium sensitive individuals can buy large numbers of paintings or other items at auctions and afterwards have a great deal of difficulty understanding why they bought items which were not even of their own usual taste. This is well known to auctioneers who prey on this type of individual. These auctioneers are hard-sell pressure types.

Lithium sensitive individuals, depending on the cycles of their illness, can be involved in varying activities to varying degrees. The binges of Lithium deficient people can involve alcohol and/or drugs. An alcoholic who is a binge drinker is very likely to be Lithium deficient. Many chronic alcohol abusers begin their drinking careers as binge drinkers, therefore it is necessary to take a careful and complete history of an individual's drinking pattern from its earliest beginnings before coming to a conclusion about a binge pattern. Alcoholism is not an illness unto itself, but rather a symptom of one of many illnesses. Lithium deficiency is one of the many possible causes.

As with binge spending, Lithium deficient individuals do not take drugs or abuse alcohol to feel good. Rather they use drugs and/or abuse alcohol to stop feeling so badly. This may be at either pole of their cycle. Lithium deficient people may drink excessively when they are "high" in order to be able to come down off the "high," or at least get some temporary rest. When they are "low" or severely depressed, they may abuse alcohol to stop feeling badly or to shield themselves from the pain from which they suffer.

Alcohol consumption is a topic of major concern and importance. When alcoholics who are Lithium deficient drink, it is like flipping a switch. They do not take one or two drinks; they cannot stop until they have a "buzz on." Lithium deficient individuals may not remember their experiences while under the influence of alcohol, and it may also cause significant personality changes. People who are usually friendly, pleasant, loving, and caring may be transformed into surly, vicious, or even physically violent individuals. They may perform acts or do things out of their usual character of which they are literally unaware. Many wives who have been physically abused by their husbands have heard their husbands vehemently deny such abuse when they are once again sober. This situation is more easily understood when the husband is found to be Lithium deficient. Lithium therapy will by no means eliminate wife and/or child abuse, but I strongly feel that there would be a significant decrease if the individuals suffering from this mineral deficiency were to be supplied with minimal amounts of Lithium Carbonate.

Let us take a close look at the "now" generation. Many parents and adult observers of today's young seem to feel that the younger generation is "going to the dogs." In most cases, this is not true. The young people of today are just as intelligent, creative, caring, and loving as we were at their age, probably even more so. Unfortunately, they are also just as sensitive to a Lithium deficiency, if not more so. Because there is a more acute shortage of Lithium in our present environment, the sensitivity to the deficiency is showing up even earlier.

As with alcohol abuse, the abuse of drugs today is also in part an attempt by young people who are even more Lithium sensitive to stop feeling so badly. Many people say the "kids" of today are just trying to escape their responsibilities. Many of these younger individuals may be Lithium deficient to such an extent that their mood inconsistency prevents them from carrying out their responsibilities in an efficient manner. Their lack of cooperation may therefore be more a matter of circumstances than a lack of concern or caring.

Since almost all Lithium sensitive individuals need peer acceptance for their emotional sustenance, the younger generation is also subject to a great deal of peer pressure to join in drug experimentation. The young people of today are subject to the "double bind" concept in drug use: if they don't join in, they are considered "straight and stuffy"; if they do join in, they often find that the drugs they are experimenting with don't give them the relief they seek. They may find temporary relief while certain drugs are active, only to find their mood swings are even more severe when the drug wears off.

Of all the drugs available today, marijuana more than any other negates the effectiveness of Lithium. In many, if not most Lithium sensitive individuals, marijuana reacts in the system as an intense, mind altering drug. Even when an individual is taking supplemental Lithium, marijuana may negate many of its beneficial effects. It is my clinical impression, based on observation and evaluation of many dozens of patients, that the intemperate use of marijuana may render Lithium Carbonate ineffective to the point where it is almost as if the individual is not taking any.

I do not mean to imply that drug abuse is common only among young people or that alcohol abuse is restricted to the "adult" community. Lithium deficient individuals of all ages are seeking relief from their emotional sensitivity and mood swings. They may seek out any means available to them.

Sexual behavior is among numerous "binge behavior" characteristics of Lithium deficient individuals. One type that is somewhat hidden, is the maintaining of multiple physically intimate relationships at the same time. This situation is a combination of "binge" because of the number of relationships, and emotional sensitivity due to the Lithium sensitive individuals' desperate need to be loved and touched. The more severe the sensitivity to the Lithium deficiency, the more likely there will be multiple relationships because of the greater need to be loved. Because it is extremely difficult for Lithium deficient individuals to maintain a consistent emotional level, they are very unlikely to find "true love." Sadly, they often find themselves trying to compensate for this by excessive sexual relationships. These relationships may appear to be intense, however they are usually superficial and fleeting.

In my clinical practice, I have treated several "nymphomaniacs." A nymphomaniac is a female who gets involved in excessive sexual activity. I have had two patients who have demonstrated nymphomania to an extreme because of their Lithium deficiency.

One woman went into a bar and offered herself to every patron there. By the time she had finished, about forty-five men had partaken of her services. After she completed this ritual, she returned home, anxious, tense, and very distraught. Fortunately, she was able to hide this behavior from her husband.

Her behavior pattern was episodic and without any obvious precipitating event. These episodes occurred during periods of relative happiness. as well as during periods of relative despair, during periods of relative success as well as failure; during periods of contentment as well as periods of disappointment.

The external lack of a clear-cut pattern is characteristic of the cycles of Lithium deficient individuals. There is a pattern although it is not obvious. The pattern is internal and has its own momentum because of the Lithium deficiency. It rears its ugly head irrespective of the external environment and it becomes an all-consuming desire. It transcends moral, ethical, and religious inhibitions. Its very nature may be contrary to the individual's basic moral code, but it nonetheless takes hold.

The woman I just described suffered anguish, sorrow, and self-contempt. She could not believe she actually did what she did, but in reality, she knows it is true.

The second female in question was a "weekend binger." For years, she was able to sustain a pattern of sexual behavior in which she would be with one individual for many hours until she exhausted the male partner, and then continue in the same manner with three or four other men. She maintained this situation on a relatively routine basis.

This "driven" individual is no less Lithium deficient than our other example. The difference is a variation on the same theme. Their relative lifestyles and environmental situations determined the outlet of their basic drives. The method of expression is secondary to the basic situation. The common denominator is Lithium deficiency.

Excesses are common with Lithium deficient individuals. They may be excessive in the use of alcohol or drugs, in sexual behavior, or something as simple as overeating or even undereating. They may be excessive in their thoughts, ideas, or feelings as well. They are impulsive binge people, literally capable of getting into their car and driving across the country on a moment's notice, especially if someone tempts or dares them.

The ability to deal comfortably with the environment is absolutely essential in order to find their place in life. The inability to find a consistent internal environment causes great pain, turmoil, and agony. This is painful for the individuals involved, but equally so for those around them. The effect on their family, friends, and loved ones is as extreme as the behavior itself is on Lithium deficient individuals.

Charisma and Manipulation

All Lithium sensitive individuals do not have all the characteristics we have described nor do they have them to the same degree. The genetic personality characteristics of each of us are modified by environment and experience. Lithium sensitive individuals who have an extra dose of verbal abilities and quick-mindedness may be said to have "charisma."

We are all familiar with people whose very presence ignites a crowd. When they enter a room, the whole mood of the room heightens and intensifies. All of us communicate on multiple levels. Communication is both verbal and nonverbal. Nonverbal communication can be as obvious and concrete as body positioning, better known as "body language." In addition, nonverbal communication can be expressed subconsciously. We have all met people who create within us an immediate positive or negative "gut" feeling about them or about our ability to establish a relationship with them. Individuals who have this poorly defined noncognitive quality of "emotional charisma" can subconsciously appeal to a large number of people in a positive manner. This positive appeal is usually effortless, natural, and a powerful motivator of others' behavior.

John F. Kennedy is a classic example. He had an appeal to the populace which transcended race, religion, sex, or age. There are many accomplished, polished, skilled, and well-rehearsed orators who do not have the capacity to get their points across to their audiences. On the other hand, there are individuals who may be less accomplished, less polished, and less precise in their command of the English language, and yet they can have throngs of people cling to their every word and movement. These rare individuals attain the greatest of all heights and leave behind the most lasting impressions.

These individuals are known to be manipulators. Most people think that word has a negative connotation, but it is very important to stress that manipulation can be, and most often is, a positive word. Parents manipulate or guide their children to do their homework. Professionals do the same to their clients or patients to get them to do what is best for them. When someone says "thank you" to the person who has held the door open, that individual is being manipulated into a relationship which would encourage the door to be held open once again. Terms such as "thank you," "would you please," "I would appreciate," "it was very nice of you," are all positive manipulators of individuals' behavior.

On the other hand, there are conscious negative manipulators known as con artists. The movie, *The Sting*, is a classic example. Con artists laid out an intricate and precise plan to manipulate or guide another individual's behavior for their own profit and gain. The precision plan also took precise individuals to carry it off. It is this ability to "feel the role" which is of utmost importance for the actor to properly act his part, i.e., for the con man to pull off the scam.

The concept of manipulation is therefore neither for nor against you, but rather it is how people utilize a manipulation that can result in a positive or negative end result. It has been said that guns don't kill people, people kill people. Understanding the personality of a con man can allow us to maximize our overall effectiveness in dealing with such an individual.

How to Con a Con Artist

The most efficient way to deal with "feelers" who are in a high mood is to let them "decide" the rules of the game. When you remove the potential "stinger" from cunning, crafty, talkative manipulators, their sense of fair play becomes obvious. It is not until there is no gain to be made from playing the "game" that the sense of fair play emerges.

If you attempt to out-manipulate mainpulators, their "wheels" begin to turn and the outcome is a time-consuming and intense battle of wits. There is a thrill beyond description when Lithium sensitive individuals become embroiled in mental gymnastics. An emotional excitement evolves which makes the consequences of their activities secondary to the actual playing of the "game."

When you refuse to get embroiled in the give-and-take of these mental gymnastics, the outcome is usually quite beneficial. Tell Lithium sensitive people with whom you are negotiating, "Why don't you prepare the contract. I'm sure it will be fair to all parties concerned," and watch what happens. In most cases, you will find your interests are well protected. The financial consequences of a contract are often secondary to the negotiation itself.

A protective sense of fair play comes to the forefront and remains resolute. Lithium sensitive individuals put on shining armor, mount the faithful steed, and protect the defenseless. The goal has become to be absolutely fair to the adversary because this one-time imaginary foe or adversary now has become a "friend in need" or a "damsel in distress."

The best and most efficient method for dealing with con artists is to allow them to outwit themselves. If you are dissatisfied with the final contract, you don't have to accept or sign it. If you do accept it, they will usually carry out their part to the fullest, and then some. They have created an image of themselves which must be upheld, irrespective of personal loss or hardship.

Giving and Receiving

Lithium sensitive individuals get much greater pleasure out of giving gifts than receiving them. They have a natural tendency and desire to shower others not only with their personal love and attention, but also the tangible gifts as well.

Lithium sensitive individuals get great pleasure out of handpicking a particular gift tailored to a particular person. Their gifts are usually unique and very individualized. They have a tendency to "overdo." They look for what they feel must be the right gift, however because they are impulse oriented and oriented to the moment, they are unable to spend a great deal of time looking for the particular present. They want to find it immediately and when they are not able to do so, they have to settle for something else. This presents another problem: when they settle, they usually purchase something even more expensive than they originally intended.

Lithium sensitive individuals frequently create a "double bind." When they give gifts, they usually try to make each gift somewhat more stupendous or expensive than the previous one. This situation creates a self-defeating spiral. Most often, the individuals are either financially or creatively unable to keep up the pace they have set for themselves. Often the givers transfer their frustration to the receiver and may treat the receiver negatively. The receiver of course is unaware of what is happening, but realizes there is a negative atmosphere in their relationship. In the extreme, givers may temporarily, or even permanently, sever the relationship because they feel guilt and anxiety when they are around the receiver. It must be remembered that the receiver is totally unaware of what had or is transpiring. One can see how a mutual, giving, and loving relationship can degenerate rapidly into a mutually hostile and negative one. When two Lithium givers are also mutual receivers, the problems become quite significant and most often destructive. I have seen both individual and even family relationships of many years standing become disrupted and fragmented when two individuals tried to outdo each other. There has always been some question about whether this is conscious or subconscious. It has been my clinical impression that at least during the early and middle phases of this mutual giving competition the motivation is subconscious. As the relationship becomes more and more uncomfortable, the search for a gift reaches a conscious level.

Strangely enough, Lithium sensitive individuals may also give gifts to individuals they don't like. Because they are extremely sensitive to rejection, if Lithium sensitive individuals meet someone who does not appear to like them, they may go to great lengths to win that person over. Still in tune with their basic personality, they first give of themselves and their time. They may flatter, flirt and volunteer to do little favors for the individual in question. They will then progress to buying items of interest for the individual, and so on and so on, until they have either "proved" worthy to this individual or have given up and withdrawn into the shell of rejection.

It would seem that individuals who are this giving would also thoroughly enjoy receiving gifts, but they can't. They have a very difficult time receiving compliments because they feel unworthy, inadequate, and have low self-esteem, and for the same reasons, have a difficult time receiving gifts. Most Lithium sensitive individual express disbelief when other people say they care for them. "Feelers" can imagine themselves caring for someone, but not the other individual returning the affection. When the other individual becomes demonstrative and gives gifts, Lithium sensitive individuals are overwhelmed and do not know how to respond. In typical fashion, they overreact to the situation by a short, embarrassed response such as a whispered "thank you" or the opposite extreme, a very verbal, gushy, overdone expression of gratitude which is out of proportion to the situation.

Even more difficult for Lithium sensitive individuals is to be asked if they would like to have a certain item as a gift. A surprise gift is hard enough to handle, but now a decision has to be made in relation to the gift. In their typically indecisive manner, they often will be unable to give a definite answer, even though they usually know full well whether or not they would like the item in question.

The hardest gift receiving question Lithium sensitive individuals may hear is, "What do you want as a gift?" "Feelers" are probably not just indecisive, but overwhelmed as well.

When Lithium sensitive individuals receive gifts, they more than likely use the first available opportunity to give something to the giver. Most Lithium sensitive individuals have a sense of wanting to be at least "even" with the individual who gave something to them. They often feel they "owe" the other individual until they have reciprocated.

I have often been asked which people are the most demanding. The answer is as complex as it is simple. It has been my clinical experience that the most demanding people are those who are at least outwardly the most giving. I recently met a young lady whose chief complaint was, "I don't understand what's happening. I am so kind to all these people, so giving, so loving, and ask so little from them, why do they leave me?"

As her story unfolded, she revealed that three days before coming to my office, she had broken up with the fourteenth boyfriend in a long and ongoing succession. She was very tearful and almost sobbed. She came across as being the victim, not the perpetrator, being taken advantage of and not wanting anything from the other person.

She then went on to describe how she would tend to the needs of her lovers and boyfriends. She described how after a relatively brief period of time she would ask her new boyfriend to come and live with her. She felt each new involvement was different: "This one will last."

People who are Lithium sensitive do not learn well from past experiences, especially when it relates to their personal or emotional life. They don't generally make the same mistake over and over, but rather a new mistake again. She described how with each new lover she would unselfishly and without hesitation wash his clothes, shine his shoes, lay out his clothes for the next day, even prearrange his toiletry items, as well as drawing his bath or shower. She described how she would effortlessly prepare multicourse meals and serve them on a linen table cloth with candles glowing, with the proper wines and wine glasses. She described how she did all of this while maintaining a forty-hour-a-week job and caring for her home. She described how dinner would be ready at the precise time each day when she knew her boy friend would be arriving home from work. The novelty of these services soon would wear off for the newfound lover. What he once greeted with open arms, he rapidly tired of and he felt entrapped and smothered. One afternoon he asked. "How about a simple hamburger tonight and some French fries?" Instead, she prepared filet mignon and baked, scalloped potatoes. He wondered why he couldn't have peas and carrots instead of asparagus with hollandaise sauce. He began to wonder what ever happened to the simple pleasures he once enjoyed, why he could not have a beer, instead of wine, with dinner. He soon recognized that when he asked these questions, she became hurt and felt "rejected." She would say, "After all, I'm only doing it for you." The excessive attentiveness which was at first a positive and warm feeling became an overwhelming frustration. He lost his sense of privacy and felt as if he could not choose his aftershave lotion according to his mood, but had to use the aftershave she wanted him to wear that day. Positive became negative; what initially was pleasure became painful.

The final straw came one day when his fellow employees asked him to join them for a round of drinks after work. He suddenly became concerned and almost panicked. He desperately wanted to join his friends, but he was concerned that if he didn't arrive at his home on time, he would feel guilty about having ruined her candlelit dinner. He felt as if he was no longer in charge of his own destiny, as if he could no longer determine what he wanted to do. The problem became more complex because it was difficult for him to get a handle on what was happening. Nothing negative was happening; in reality, his life had become an uncomfortable negative. After all, how could he get angry with someone for being too attentive, too caring, too giving, too loving? How? It's easy when the caring becomes painful. One day he went home and told her, "Things are just not working out. I don't know why. I don't know what to say. I don't know how to say it. I just know I'm not comfortable here anymore." A few hours later he was gone. She could not understand what she had done wrong. In fact, she could not comprehend that anything was wrong. She was giving; he was taking. She was loving, he was uncaring. She was unselfish; he was selfish. She was kind, he was unkind.

As she continued to tell me her story, she appeared perplexed and asked how someone as willing to give as she could meet so many people who took advantage of her and then left. As she was leaving her first session with me she said she intended to be much more careful and selective before she would allow herself to get involved with another man again.

When she returned to my office two weeks later, she told me how fortunate she was that she finally met the one man she knew she could and would make happy. She said she had learned her lesson from her other encounters and certainly would not rush into having number fifteen move in with her until she knew him for at least another week or two. She told me that she was no longer depressed, that she didn't need psychotherapy, and that her current romance would enable her to readjust to her daily routine and pressures.

I tried to get her to recognize that perhaps this relationship was forming too quickly and that the end result might be the same as the previous fourteen. She was sure I was wrong. As the session ended, I told her that if by some chance this relationship did not work out and if she were then ready for a meaningful exploration of her behavior, I would be happy to make time for her. She assured me this would not be necessary, but if it would make me feel better, she would remember my offer.

Emotional Chameleon

The chameleon is one of nature's creatures which has the adaptive capacity to change colors according to the environment in which it finds itself. This quality has helped these creatures survive in hostile environments.

Most Lithium deficient individuals are emotional chameleons. They have the adaptive quality of being able to conform to the thoughts, feelings, ideas, and philosophies of the surroundings in which they find themselves. This adaptive capability allows them to "fit in." They are often known as "as if" individuals: they act and feel "as if" they are an integral member of the group in which they find themselves. This chameleon quality allows them to be caught up in the mood of their environment. They have an urgent desire and need to please people. They may find themselves at different times of the day or on different days involved with groups that are diametrically opposed to one another, either politically or philosophically. They often try to please both groups. Imagine a chameleon being placed on a piece of scotch plaid fabric, attempting to adapt to each and every color it finds beneath it as it walks across this piece of cloth. If it is on two colors at the same time, imagine the confusion, dismay, and physically overwhelming position in which it finds itself. With Lithium sensitive individuals, their attempts to please the large number of groups in which they find themselves become self-defeating, exhausting, and, at times, emotionally crippling.

These individuals are so emotionally sensitive and adaptable that they often become prey to cults. People wonder why college kids and children from intelligent families are the ones most easily preyed upon by these cults and splinter groups. Intelligence, emotional lability, emotional sensitivity, and many other factors, including being easily hypnotized, are genetically inherited, paired characteristics. Consequently, the more intelligent, the more easily persuaded, ingratiated, and conned. Con men prey most heavily on Lithium sensitive individuals. Individuals who reach higher socioeconomic and educational levels are more likely to have the funds which cults need to exist. As chameleon types, they depict many adaptive qualities. Being a chameleon is not in itself either a positive or a negative attribute. It is the way in which individuals use this capacity of Lithium sensitive individuals which is for or against them that becomes the major issue.

Statements like, "He wasn't a bad kid until he found himself with this new group of friends," are especially applicable to Lithium sensitive individuals. Conversely, placing individuals in a healthy, loving, giving, caring, feeling environment can override many previous negative situations. This is especially important when dealing with the concept of treatment or management of Lithium deficient individuals. They must be helped to replace their Lithium and to learn assertiveness techniques, but it is equally important to help them structure the environment around them to meet their needs so the gains they make will be lasting and enjoyable.

It is important that the full explanation of the chameleon quality of Lithium sensitive individuals be understood so their internal and external environments can be appropriately altered to meet their individual needs.

Assertiveness Training Doomed to Failure

Assertiveness training has been highly successful and rewarding, for those people who give the training courses. As with any new technique or procedure, there is always a flurry of followers at the beginning. Those who joined the bandwagon initially and then somewhat modified the original techniques have written innumerable books, given innumerable courses, and made significant profits. The overall effectiveness of all of this training is, at best, questionable. It has been my clinical impression that the vast majority of people who take these courses are unable to successfully carry out the basic tenets and principles necessary to derive the benefits they desire.

In theory, self-help courses and self-help books are very reasonable and practical. They are "obviously" going to work. Their simplicity and practicality are without question. I am not deriding the theory of assertiveness training nor its usefulness. My point is that regardless of how useful a technique is, it is still essentially worthless if it cannot be applied consistently.

The initial enthusiasm of Lithium sensitive individuals is in itself a sufficient impetus to create a positive atmosphere for almost any new concept, project, or system in which they wish to engage. The initial enthusiasm of Lithium sensitive individuals is of such magnitude as to be able to literally infuse the people around them with a similar degree of enthusiasm.

Lithium sensitive "feelers" are impulsive, explosive, grandiose, and excitable. When the "go button" on the "feelers'" computer is pressed, they begin a new concept or project with enormous amounts of enthusiasm and excitement. They easily adapt to a new project or system and assimilate the new information as a sponge absorbs water. They embrace all the facts and then begin to proselytize the concept or system. They feel that anything that they are interested in should be of interest to the whole world. Since they have chosen this concept for this present moment for themselves, so should everyone else. "Feelers'" enthusiasm is so great as to have an infectious effect on the people around them. When Lithium sensitive people accept a new concept it is then an integral part of them as a person. For another individual not to utilize the system is, therefore, a direct insult to their very being.

"Feelers'" enthusiasm is all-encompassing, at least for as long as it lasts. As with most other projects, Lithium sensitive individuals begin to lose enthusiasm as they develop enthusiasm for another concept, project, or system.

Assertiveness training requires absolute, or almost absolute, consistency in dealing with yourself and with another individual. When a child spills a glass of milk on Monday and the parent reacts by screaming at the "careless" child, the child develops a certain anxiety level in relation to what will happen if he or she spills a glass of milk. The level of expectation of the parent's behavior has been created in the mind of the child.

On Tuesday, the same situation may result in the parent silently cleaning up the milk. On Wednesday, if the child spills a glass of milk, the parent may suggest they clean it up together. By Thursday when the child spills his glass of milk, he is totally overwhelmed, anxious, and confused about how the parent might react. As simplistic as this may seem, it could be applied to all interactions between individuals on any scale or intensity of situations. Consistent reaction is the most crucial and essentially the all-important factor in any relationship. It is absolutely necessary for a child to be able to trust his or her parents. It is absolutely necessary for spouse to be able to trust spouse. It is absolutely necessary for an adult to be able to trust his or her child. Trust is essential for a firm foundation on which to build any relationship. It requires a consistent approach to a problem so an individual can understand the motivation of another person.

Without consistency, there is no trust. If you do not know how you would most likely react to a given situation, you are unable to even trust yourself. If you are unable to trust yourself because your reactions are inconsistent, others will also be unable to trust them.

Lithium sensitive individuals are literally carried along with the tide. They react to the mood of the group with which they are associated and can therefore interact with many types of individuals whose expectations of their behavior may vary greatly. Their need to be accepted by their peers governs their attitude and behavior. They lack a consistent approach to situations. Their very being dictates that they are consistently inconsistent. Their needs are subject to the expectations of those with whom they find themselves involved.

Despite an individual's ability to learn and hopefully utilize assertiveness techniques, no one can use them consistently unless they are consistent within themselves. It is most unusual for individuals to be at their peak all the time. Normal mood variations are expected and anticipated. It is the overwhelming, overriding, intense mood swings however which cause an inability to deal consistently with others and with oneself.

We return to basics: individuals must be consistent in their relationship to themselves and to others or they cannot be successful either as individuals or in their relationships to others.

The major factor in establishing internal consistency is adequate levels of Lithium Carbonate. When individuals develop this consistency, they feel better about themselves. They act, they think, they make decisions. They become open and honest. They read situations as they exist, rather than how they want to see them. Their rose-colored glasses become clear . . . reality prevails.

Chapter Five

WHY IS IT YOU GROW, BUT DON'T CHANGE?

In classic psychiatric literature, manic-depressives (Lithium sensitive individuals) usually become ill in their late thirties or forties. The description stated that individuals suffering from this illness exhibited classic symptoms of intense depression followed by, or interspersed with, euphoric, happy, elated highs. They were described as having grandiose ideas that made them think and feel as if they could walk on water and/or fly without wings. Classically, as the illness progressed, depressive features became more frequent and more prominent. These depressions would be so severe that hospitalization and often electroshock therepy were necessary. The precursor personality was known as "cyclothymic" or moody and sensitive.

There has been an important change in the natural progress in understanding and detecting this most common illness. It has been my clinical impression and understanding that individuals who are genetically endowed with a sensitivity to Lithium deficiency can be picked up even in utero (still in the mother's uterus). Most people suffering from this sensitivity have been described by their mothers as unusually active and frisky fetuses. They report their offspring kicked early and at a most unusual rate. As you will remember from previous chapters, Lithium Carbonate, as a natural mineral, has been effectively removed from our water supply and, therefore, is not being replenished in our bodies. With each successive generation, females have had less and less Lithium to share with their offspring. This also accounts for why younger children in a family are more apt to suffer from a Lithium deficiency than are their older siblings. As each child is born, the mother has less and less Lithium to give. In a family of four Lithium sensitive children, the oldest will usually have the most Lithium, and the youngest the least.

Females who have a sensitivity to Lithium deficiency, or who have the overt characteristics of Lithium sensitive individuals, will in all likelihood produce offspring with a greater degree of Lithium deficiency because they have even less to share. These offspring, while in utero, show greater and greater activity and mobility.

It must also be stressed that when Lithium sensitive mothers give birth and give up some of their Lithium reserve, they often goes through a period of significant and severe depression. Medically, this situation is called postpartum depression, also known as the "baby blues."

Postpartum depression is unique to women. I sometimes feel that the medical profession does not investigate with the same proficiency illnesses or situations that are unique to women as illnesses indigenous to men. Postpartum depression is an overwhelming illness to those unfortunate women who are afflicted with it. After childbirth, their lives become nightmarish. At times, many of them are literally unable to function. They can not take care of their child's needs and often are unable to even care for themselves. This illness can become severe enough to require hospitalization and/or the presence of a caretaker in the home.

Lithium Carbonate is a mineral which unfortunately is not replaced by standard vitamins or mineral supplements. Prenatal vitamins do not contain Lithium. As the fetus extracts Lithium from the mother, her supply when the child is born is significantly and dramatically diminished. This decrease, in my opinion, is responsible for most cases of postpartum depression. The consequences of acute Lithium loss are profound. It immobilizes individuals at risk and causes overwhelming emotional consequences.

I have learned by studying women who have had postpartum depressions that in most cases there is a positive family history for the existence of a circular mood disorder. The chance of females developing postpartum depression significantly increases with positive family history. Additionally, if they are moody, notably sensitive, and/or have a history of frequent or relatively frequent depression, chances of a postpartum depression increase significantly. This increase is of such amplitude as to indicate the need for the prophylactic use of Lithium Carbonate. Based on clinical evidence and observation, the use of Lithium Carbonate in the eighth and ninth months of pregnancy, or just in the ninth, could significantly reduce the possibility of developing postpartum depression. It has been my clinical impression that should these mothers be offered Lithium Carbonate even after delivery, their depression resolves rapidly.

Lithium deficient newborns often show unusual and extreme behavior patterns. They are often poor sleepers, may take infrequent naps, are usually extremely demanding and frustrating to deal with. Lithium sensitive children are unusually active in the nursery. They often have little red marks on their knees from trying to push themselves around the crib. They feed relatively poorly and are awake a good deal of the time. Conversely, the opposite extreme can also exist. Children who are Lithium deficient may instead show little activity and be relatively quiet, sleepy babies. The operative word is "extreme." Lithium deficient individuals are, in fact, recognized by the extremes of their early childhood behavior. As you will remember, a basic characteristic of Lithium sensitive "feelers" is that they are very tactile people, very kissy, huggy, and touchy. They need to hold and to be held. Lithium deficient infants demand to be held, and if they are not, they frequently cry and fuss.

They are often very poor eaters. They do not concentrate well on the bottle or the nipple. They lose patience and become whiny and irritating. This situation is further complicated since Lithium sensitive mothers who have just given birth, are at a low ebb in tolerance because of recent radical Lithium loss. Consequently, we often see a highly frustrated mother trying to deal with an extremely demanding child. This often creates a situation which can give rise to child abuse.

When a mother with a low frustration tolerance has unusual and extreme demands placed on her, she justs wants to be left alone. Since she is impulsive and now oriented, her natural tendency is to use whatever means are at hand to cope with the situation. Physical violence and/or extreme verbal abuse are unfortunately expedient and ever present. The extreme of the abuse is in proportion to the mother's tolerance level. She does not consider the consequences to herself or the infant. The consequences are, however, an ever present concern of society in general.

As Lithium sensitive children begin to reach their early milestones, they usually do so at a rapid rate. Children who are born with a sensitivity to Lithium deficiency are very bright and progressive. They pick up their heads, sit up, stand, and walk early. They are usually verbal and learn to speak at a young age. At the same time, they are demanding and often known as "gimme" children. They are impatient and impulsive. They want and demand that their needs be met instantly and, yet, they change their demands frequently.

As they continue to grow, they seem to have only two gears: stop and go. When they sleep, they fall into a state of exhaustion. They sleep soundly and as intensely as they do everything else. When they are awake, they are like whirling dervishes. They move about frequently and untiringly. It is physically and emotionally exhausting to deal and be with them. They demand and exact a great deal of time and caring from their parents. The younger children are usually most Lithium deficient and when their heavier demands are placed on the parents, it is usually in conjuntion with many other demands being placed on them by older children in the family. Each successive child appears to put more strain on the family unit. Most likely it is not just the addition to the family of another person, but the increasing Lithium deficiency of each successive child.

As children continue to mature and approach the preschool years, a new set of problems may develop. Since they are usually quite intelligent and verbal, they are likely to develop a habit of incessant questioning. They are initially viewed as cute and precocious, however this soon changes to being irritating and exasperating. They ask questions over and over and over. When they are given an answer, they begin to question it. Their verbal abilities are a double-edged sword.

We now have passed through the stages of overactive physical activity, compounded by overactive verbal activity, better known as motor-mouth and motor-legs.

The exquisite sensitivity to rejection shows itself in preschool and early school activities. These are individuals who either share well in order not to be rejected, or refuse to share at all because their toys and possessions are part of them. Keep in mind the variety of extremes these individuals show. One moment they may share well, and the very next moment refuse to share at all. At one moment, they may grab and be very demanding. The next moment, they might be very loving, giving, sharing, and caring. The frequent moodiness and changeability of their behavior is part of the most basic characteristic and diagnostic criteria of individuals who suffer from a sensitivity to Lithium deficiency. They might want to sit on your lap, hug you, kiss you, and touch you, and a few moments later, they may want nothing to do with you at all and become very rejecting. They can comfortably play by themselves for hours when they are in the "mood" to do so, or they may refuse to be left alone.

When they are willing, they can easily be taught arts and crafts and even music appreciation at a very young age. At other times, they can be obstructionistic, overly playful, and destructive to a learning program. They can play well with others, or they can be impossible. Their moodiness creates within their parents and siblings a generalized feeling of disquiet. This feeling exists because individuals who deal with moody individuals become unclear about what to say or how to say it, what to do or how to do it. They are subconsciously attempting to respond to the ambiguous demands of a "high."

As children approach school age and enter school, a whole new myriad of stresses becomes apparent. As they enter school, thoughts of rejection (that is, being sent out of the home to school) and the difficulties experienced in dealing with a foreign and radical change in environment, become apparent. The initial adjustment to kindergarten and/or first grade is often tumultuous. There is an exquisite sense of being "rejected by the parents" and being sent into the school system "alone," and an intrinsic and obvious concern about whether or not they will be accepted by their peers. Peer acceptance is one of the most crucial aspects of a healthy education experience for individuals who suffer from a sensitivity to Lithium deficiency. A lack of peer acceptance has dramatic effects on all aspects of the ability to become homogeneous members of the new community, to have a positive educational experiences, and to be able to engage in extracurricular activities.

I cannot stress too strongly the absolute and almost overwhelming negative effect which occurs in Lithium sensitive individuals who lack full acceptance by their peers. These individuals are exquisitely sensitive to any overt signs of rejection, and they frequently internalize and misinterpret external events which make rejection even more prominent.

When they review the result of miscommunication, Lithium sensitive individuals do not say, "What is wrong with this situation?" but rather, "What did I do wrong?" "Feeler" children, as well as "feeler" adults, tend to take "blame" for any situation that does not work out as expected. This places additional stress and strain on these individuals and causes an even more significant amount of energy to be spent coping with the routines of daily school attendance.

It is important to keep in mind that every individual is born with a certain amount of psychological energy, but only a relative amount of psychological energy to help him get through each day. If an individual spends most of the psychological energy just trying to cope with the routines of daily living, there is very little energy left to appreciate the world and enjoy the things and people in it. Consequently, Lithium sensitive individuals tire quickly as the day goes on. There is such a vast expenditure of energy that crankiness, unstable behavior, rebelliousness and sheer orneriness usually increase as it gets later. When educators fully understand this, they will make changes in the curriculum to maximize the intensiveness of educational experience in the early part of the day, leaving the later part for recreation.

Lithium sensitive individuals do not adjust well to change. Once they settle into a given climate or environment, they want to stay put. Therefore, it is not difficult to understand that a child who does not want to leave school on Friday afternoon may not really want to go back to school on Monday morning.

The theme that seems most prominent about the sensitivity to Lithium deficiency is that the variables are the only constant. A circular mood disorder is a sometimes illness. Individuals can go through periods when they seem to be relatively quiescent and peaceful, and then suddenly have significant periods of emotional upheaval.

This situation becomes even more complicated when children who suffer from these mood changes are involved in either Child Study Teams and/or private therapy. When individuals who are in therapy change, everyone assumes the change has to do with therapy. Unfortunately, this is not always, or even usually, the case. Often the changes occur irrespective of therapy, because of the "normal cycles" of the mood swings themselves.

As young children continue to progress through these years in school, they may either go totally unnoticed, depressed, withdrawn, silent, literally a quiet bump on a log, or the opposite may be true. The "squeaky wheel gets the grease," so youngsters who disrupt their classrooms by behaving badly are noticed and more often are stringently dealt with. Acting out impulsive, explosive behavior catches the teacher's eye and makes it most likely for these children to become involved with "child study teams."

During these early stages of education, teachers must become more acutely aware and cognizant of the psychological needs of their students. These teachers are in a most advantageous position to recognize youngsters who have exquisite sensitivity, whose feelings are "easily hurt," who would rather not risk rejection and therefore withdraw into their own private, silent world. These individuals are behaviorally the "no problem children," and their psychological needs often go unnoticed. They are quiet and pleasant, happy to clean the blackboard and erasers after school. They volunteer for extra work to get an extra measure of acceptance. As they become more accepted by the teacher and are praised for their willingness to help they become targets for criticism by their peers. The "feelers' " double bind situation applies even to young "feeling" children.

The Shirley Temple Syndrome

Children whose parents are visceral, emotional, and have the sensitive vibrations of "feelers," will in all likelihood have the same or similar characteristics. Most sensitive children are rather coquettish, actually seductive, and at times are downright alluring. They control by their cuteness and are demanding by being enticing.

Lithium deficient children are so likeable that people don't get overly annoyed with their controlling nastiness. They are nasty in a pleasant way. They interrupt, demand time and attention, and can monopolize a conversation or a household.

The Daydreamer

Many children come home from school with notes that say, "Your child is unusually bright, but he does not seem to be able to apply himself." Or, "Your child does not apply himself and obviously is lazy." Or, "We expected much greater things from him, but he doesn't seem to live up to his potential."

What is happening? Why are so many children doing so poorly? Why are they underachievers? Why do teachers write, "Your child just seems to sit and stare out the window and doesn't really seem to want to pay attention"? What is happening is that we have children who, because they usually have bright parents, are genetically endowed with more than average intelligence and are to suffer from a sensitivity to a Lithium deficiency. The children of the gifted, who are themselves gifted, often suffer from this sensitivity syndrome. They are bright, alert, intelligent, imaginative, intuitive, sensitive, either shy or unusually outgoing, but because of their innate capacity for the daydream or spontaneous trance state, they appear to perform below their potential.

This subpar performance level is itself not a conscious choice. These children do not intend to daydream; it is a natural phenomenon. The daydream state and the hypnotic state are either self-induced or are induced by another person; the daydream state is spontaneous. Individuals are paying attention, but a certain part of a lesson or statement triggers a fantasy in which they become wholly involved. A similar situation may occur when children become temporarily bored by something in class, which triggers the initiation of a daydream. A daydream may be triggered when noise or something happening outside the classroom momentarily distracts. Individuals who have the genetic capacity that allows multiple daydream states to be easily triggered often appear not to care, but the opposite is true. Most children who daydream spontaneously can't help it. Because of genetic endowment well beyond their control, average individuals who daydream do so almost totally without knowing it. In fact they are often concerned and disturbed by the fact that they daydream and that no one cares. They recognize that daydreaming represents a very significant problem for them and yet they are unclear about what to do about it.

The concept of daydreaming as involuntary and the fact that others do not pay attention are very important. This situation is compounded when parents further punish or admonish children for a situation which is out of their control. The children may then develop hostility and resentment toward their teachers and their parents. This feeling may become compounded further if something interferes with their free time and play time since Lithium deficient children usually have very high energy levels and need play time to work off some of this energy.

Children who are gifted are also cursed. It is a very frustrating and overwhelmingly irritating when bright, intelligent children are unable to perform to the extent of their abilities. This frustration and anger leads to many types of personality disorders.

The types of personality disorders that develop are as varied as the individuals. They may become unusually shy, withdrawn, and retiring, to avoid being questioned in class and, therefore, to avoid criticism. A second group become the class clowns. They are boisterous, annoying, and offensive to the teacher. They are verbal, abusive, and disconcerting, and often simultaneously cause hostility and laughter among classmates. They frequently challenge the teacher and are unwilling or unable to respond to the requirements of classroom decorum. They are impervious to insults, do not learn from experience, repeat the same mistakes frequently, and are often sent to the principal's office.

They avoid criticism by not letting anybody put them on the defensive. When criticism is levied, they view it as sour grapes on the part of the other individual. They are unwilling and/or unable to deal with the consequences of their behavior. Most observers view these children as incorrigible, because they are verbal and antagonistic. They receive a lot of attention. Although this attention is negative, it is still attention. They require a great deal of love and caring. Verbally and nonverbally, they demand caring and attention from their teachers, their teachers' superiors, and their parents. Although they are criticized and punished a great deal, this input is better than no input at all.

Most parents overpunish and rescind. The child is punished, but the punishment usually is not carried out. This lack of follow-through and inconsistency in the way they are handled further complicates an already complex situation.

Quiet, shy daydreamers are basically ignored. Because they don't exhibit behavior problems, they do not come to the forefront of classroom problems. They remain withdrawn and retiring and are not noticed. They form few friendships and are usually social outcasts. This group of Lithium sensitive children respond to the theory that although it is very painful to be alone, it is better to be alone than to be rejected. At times, they will not even answer questions that they know the answers to because they are afraid their peers will think they are too smart and not befriend them. They are oversensitized to criticism and ridicule, and allow others great freedom in the way they allow themselves to be treated, rather than subject themselves to the possibility of rejection. This concept is always at work in the minds of Lithium sensitive individuals. Their needs and wants are not out of keeping with the main flow of society, but they perceive themselves through a chronic haze of helplessness, insecurity, inferiority, and worthlessness. They do not even want to be tutored because they feel they are worthless.

Although daydreamers are genetically bright, intuitive, and creative, behaviorally they come across quite the opposite. Children who can do very well on verbal intelligence tests but poorly on written ones should be evaluated to see whether or not they are plagued by Lithium deficiency. Children whose scores vary greatly on different tests at different times should be evaluated in detail to learn if the testing difference comes from mood variation. The lives of children who are Lithium deficient can be altered dramatically by using moderate amounts of Lithium Carbonate. It is easily, efficiently, and inexpensively administered, yet it could save a school career.

The Hyperactive Child

Hyperactivity and hyperactive children have been investigated widely within the last ten years. These children are categorized as having anything from minimal brain damage to neurological impairment to minimal brain dysfunction to hyperactivity, etc. There are dozens of labels which have been applied to these individuals, and there are many centers in major cities and hospitals specifically designed to diagnose and manage these unfortunate individuals.

I have evaluated a large number of "hyperactive children." It has been my impression that at least twenty-five percent of those who are diagnosed as hyperactive or minimally brain damaged, etc., are in fact Lithium deficient. Their diagnosis and treatment program is essentially erroneous. The treatment is erroneous because the diagnosis is unfounded. More careful evaluation is essential in order to properly treat the children in question. In the significant percentage of individuals whose "hyperactivity" is not primarily a result of brain injury or central nervous system dysfunction, Lithium Carbonate appears to me to be the treatment of choice. Further, it is my impression, and let me make this clear from the start, that truly hyperactive children benefit from approaches that are currently in vogue. I am referring to children who do not belong in the category of hyperactivity.

A hyperactive child, by definition, cannot easily maintain a quiescent position, if at all. Hyperactive children are literally on the go all the time. They have two gears, stop and go. When they are in "go," they do so without rest or respite. They even eat and drink on the run. They have difficulty falling asleep and/or require minimal sleep.

Lithium deficient children have many similar characteristics. They are very, very active, and have a great deal of energy. They literally wake up in fourth gear, quickly shift into fifth, and stay there almost continuously. There are some exceptions. When Lithium deficient children watch a television program they chose and like, they can watch as if in a trance. You may even have to call their names more than once to get their attention. When they lose interest or the program is over, they again assume fourth or fifth gear and begin their pacing, racing activities.

These Lithium sensitive children may have many toys out at one time. They do not learn from punishment or from history. They appear to repeat the same mistakes over and over. They make a "new mistake again." When alone, they are unwilling or unable to carry out tasks like cleaning up, yet, if you were to stay with them, or even just stop back every few minutes to tell them they're doing a nice job, they can then clean up and take care of their toys. Lithium deficient children need another human vibration in their surroundings as much as Lithium sensitive adults do if not more so.

Children who are Lithium deficient usually have rich fantasy lives. They exaggerate and make up stories. They can play with one toy for hours if they choose, but if they choose not to or if their illness is in high gear, their activity is absolutely inordinate, frustrating, and, at best, exasperating to their parents.

Unfortunately, the psychiatric literature, as well as the *Physicians' Desk Reference*, or *P.D.R.*, do not have adequate information with reference to the use of Lithium Carbonate below age twelve. They are neither in favor or against it; they just state that it has not been fully investigated. Despite this drawback, it has been my clinical impression that the usefulness of Lithium in these children has been so overwhelmingly positive and so simple to maintain that I have prescribed Lithium for children as young as twenty-one months. The results have been so rewarding and so positive that they literally changed the lives of the children and the families.

It has been my conclusion and understanding that it is most difficult to get the mainstream of medicine to accept the most simplistic approaches. It is also my impression that time-honored techniques, no matter how inferior, are held onto with great tenacity and absurdity. When a clinic is set up to specifically diagnose and handle a particular illness, that illness becomes so much a part of the daily activity there that sometimes it is difficult for them to see the forest for the trees.

The Adolescent State of Development

As a psychiatrist, I believe that most, if not all, basic issues of conflict can be rendered down to a rather simplistic end. This usually consists of one or two basic options. Choice remains the province of each and every individual, however when one limits the available choices, decision making becomes relatively easy and efficient. It is my clinical judgment that all child rearing processes, if not all interpersonal relationships, can be condensed to two words: "love" and "limits." Children, friends, and family will allow an individual to make many mistakes and errors in judgment if they are convinced and feel the individual "loves" them. With each interpersonal relationship, especially in child rearing, a set of firm, consistent, but reasonable limits must be established, though, so every one in the family unit knows what kind of behavior is expected and at what point they may overreach their boundaries.

This concept of love and limits is not unique, but it is readily applied. When proper limits are not set, it causes a great deal of confusion and misunderstanding, and has allowed impulsive behavior to run rampant in our country. Individuals who are Lithium deficient are at best impulsive and indecisive and avoid decision making. They are subject to peer pressure and to an overwhelming need and desire to please. Unless firm, consistent, and reasonable limits are established for their behavior, they act according to whatever their peer and/or parental group at large expects.

Setting appropriate, functional, and reasonable limits must be established and adhered to on a consistent basis from the moment children are born. If parents wait until adolescence sets its own pace, they have waited too long. Parents should be parents and not their children's friends. They should be able to maintain a role which although complex, is well defined. Indecisiveness and lack of consistent parenting result in inconsistent expectations of an adolescent's behavior pattern. Since it is my firm conviction that the vast majority of individuals in our society are Lithium deficient to at least some degree, I think it is obvious that the current generation of parents has difficulty setting firm and consistent limits. This difficulty is understandable, but nonetheless is destructive to this generation of adolescents. When an inconsistent parent deals with an inconsistent adolescent, there is turmoil. Trust means consistency over time. Without consistency in adolescent and parent, a situation and atmosphere in which trust can be established becomes difficult if not absent. The judicious application of Lithium Carbonate to the vast majority of the population will allow consistency to be established and, therefore, trust to exist.

Lithium Carbonate, the consistency factor, is absolutely necessary to the establishment of a homogeneous and stable family environment. The lack of Lithium Carbonate causes most family conflicts and disunion. The end result is a disruption of the American family as a viable, interacting unit.

As we follow the development of Lithium deficient individuals, there are two major possibilities or extremes in their method of interacting. One example has been called the "adolescent discontrol syndrome." At the opposite extreme is the quiescent, shy loner.

The Adolescent Discontrol Syndrome

I have alluded earlier to a generation reversal. The adolescent in the vast majority of families, has become the power-to-be in the family structure. Adolescence is, as always, an extremely painful, emotional period. It is a time during which average males and females have to decide on their future vocations and have to contend with sexual development and sexuality on both individual and heterosexual bases. They fight between feelings of dependency and independence. They solidify their own self-concept and attempt to deal with personal, philosophical, religious, and political ideas and identities. The adolescent period is one of absolute turmoil, where inconsistencies run rampant. Average adolescents suffer greatly from periods of massive mood swings. They normally change their feelings, ideas, and interactions with peers, siblings, and parents as frequently as they change their clothes, if not more often. Frequent mood changes has been time-honored as "normal" for adolescence, but I am unclear about whether this is always a normal phase or whether it may be accentuated by Lithium deficiency. As this complex situation continues to evolve, it becomes further apparent that the concept of Lithium deficiency merely accentuates moodiness and ill-tempered adolescent attitudes rather than creating a set of attitudes.

An adolescent discontrol syndrome comprises all the above, plus many additional features. Tempers become even more labile; they become highly impatient, argumentative, and stubborn; they are extremely intense. This is so great that it creates an atmosphere of extreme emotion within the entire family structure. Adolescents involved in this phenomena are basically demanding, controlling, hostile and evasive, and they literally use terror tactics to run the family. These tactics are so intense that they mitigate against logical discussions.

Adolescents embroiled in an adolescent discontrol syndrome are literally impossible to live or deal with. Within the framework of individuals suffering from this syndrome there is very little latitude for understanding or compromise. Adolescence is, at best, one of the most painful transition periods of life. Its stresses do not fall just on the individual, but on the family as a whole.

The Adolescent, Isolated, Quiescent Loner

Adolescents who suffer from a Lithium deficiency and become passive, dependent loners suffer as intensely as adolescents who have developed the potential of becoming aggressive, acting-out, hostile, and insubordinate.

In school, they remain quiet, withdrawn, and aloof and it is difficult to get them to interact with their peers. They keep to themselves. They avoid criticism and avoid making friends for fear of losing them. They remain almost unrecognized by teachers and classmates.

At home, they keep to themselves, too. The fear of rejection vastly outweighs the concerns of isolation. They are nonetheless individuals at risk. Their willingness to read and study is socially acceptable so their difficulties and plight are not as readily recognizable those of their counterparts, the aggressive, talkative, openly hostile individuals.

These individuals have marginal self-esteem, a minimal sense of self-assuredness and internal security. They avoid criticism (rejection) at all costs. They are usually passive, compliant, and are considered "good" children because they create no noticeable or overt family problems. Their parents are called to school infrequently, and they are regarded as acceptable to the overall social environment in which they find themselves.

After examining both extremes, one must recognize that a significant number of adolescents are in the middle. It is my understanding from clinical practice that human behavior covers the full spectrum of Bell's curve of normal distribution. When you examine Bell's curve, which is shaped like a bell, there are a significant number of individuals who lie at both extremes, and there are also a good number of people who lie somewhere in the middle. These individuals maintain properties of both extremes, and depending on their personal "moods" and/or the "mood" of the environment in which they find themselves, each and any of the aforementioned types of behavior can be identified. For the most part, they interact in a socially acceptable manner. At times, they do exhibit extreme behavior in one direction or the other. These individuals are felt to be more open, honest, assertive, and variable in their reactions to environmental stresses. Their internal milieu, i.e. feelings of appreciation of their external surroundings, varies according to the intensity of the surroundings.

As these individuals leave adolescence and enter young adulthood, and eventually adult life, they carry with them the same feeling tones and mechanisms of emotional interaction and reaction. Adult life is merely a continuation of the feeling tone established during childhood and adolescence. The behavior of these individuals will vary according to the environment in which they find themselves and how their behavior is appreciated by others.

The ability to modify behavior is a function of motivation. I define motivation as "what's in it for me?" When an individual asks a friend, "Would you like to help me take out the garbage?" there is a certain level of response which is rather obvious. However should one individual ask another, "Would you like to go to a World Series game with me? I have an extra ticket," the level of response would usually be quite different, more compliant or intense and/or rapid.

You can readily see that adolescent and adult behaviors are merely part of a continuum of a behavior pattern which, to my way of thinking and clinical observation, is an outgrowth of how individuals handle their own Lithium deficiency. We recognize that this is once again a rather simplistic understanding and explanation of human behavior. However after treating thousands of individuals and having many hundreds on Lithium Carbonate, the clinical results speak for themselves as verification of these postulates.

All theories and hypotheses are questionable and at best exercises in mental gymnastics unless they are useful. The most important aspect of understanding Lithium deficiency is that not only is it simplistic, but that with the judicious application of Lithium Carbonate, one can see dramatic changes in the behavior of individuals at risk. The mainstay of these theoretical discussions is that each and every one of the negative aspects of the way individuals behave and interact with others can readily and easily be brought under control.

It must be pointed out repeatedly that the judicious application of Lithium Carbonate does not make individuals into robots. It does not make them androids nor are they controlled. Rather, it allows individuals to regain cognitive thought control over their own emotional or physical state. It allows them to act rather than react, to have a delay period between the time a question is asked and the answer delivered. It allows them to think about the consequences of their behavior and/or responses before they act or speak. I think the judicious use and application of Lithium Carbonate, is an absolute necessity for the vast majority of mentally or emotionally ill or disturbed individuals, and also for the general population at large, to help them deal with their feelings and emotional sensitivities.

Chapter Six

ARE YOU GETTING YOUR MONEY'S WORTH?

This next concept will have many references to psychiatric treatment. In itself, it may be somewhat controversial but it needs to be said.

It can be very difficult to understand the efficacy of many classic psychological and psychiatric techniques. Throughout my years of training, and attending numerous lectures and seminars, I learned that "tincture of time" was a major psychiatric technique. Although no one admits it, it is apparent that when an individual is seen in therapy for an extended period of time, why change occurs is never really clear. You can never be sure whether it is a function of time and life's circumstances, or whether the therapist had any influence. This concept may seem heretical, especially in light of the books already written to logically explain the efficacy of these therapeutic techniques. These volumes clearly vindicate practitioners of the mental health professions, or do they?

It is obvious that over a five-year period an individual will change. It is even more obvious that over a one- to two-year period, or perhaps even a one- to three-month period, an adolescent or young adult goes through a great deal of change. It has been my impression that the length of time an individual is seen in therapy and the frequency of visits is more a function of the individual's pocketbook or insurance program than a function of the need of the individual patient, per se.

This becomes even more apparent when people are inpatients in the psychiatric sections of a general hospital or in a psychiatric facility. The length of time that treatment seems to take is often a direct function of the type of insurance coverage they have. If individuals have "good programs," their length of hospitalization may be many months. Should their insurance program allow only a minimal stay, minimal treatment is then "indicated." These statements do not apply to all practitioners of psychiatry. They do, however, apply to a significant number to warrant exposure and discussion. Hospital utilization committees and peer review groups can readily attest to the fact that certain individuals command a much greater length of patient stay than others.

Socioeconomic strata education also significantly affects the duration and intensity of therapy. The more educated individuals expect therapy to be an involved procedure which should take a long period of time and be emotionally painful. It is difficult for them to relate to the fact that therapy can be short-lived and intensive. Their expectations are a function of their own educational background and exposure to the system as it exists.

A clear-cut example of this follows. I was privileged to meet a young gentleman who professed to have gained a great deal of insight and understanding into his relationship with his mother through five years of therapy. During these five years, he saw his therapist three times a week. At the beginning he found his mother a very hateful, despicable object whom he abhorred and avoided. Towards the end of his fourth year in therapy, his mother died and left him three million dollars. Shortly after, he explained to his therapist how successful therapy was, and that he now thought fondly of his deceased mother, and in a positive vein restructured his memories of her and his childhood. Tincture of time, the therapist waiting four years (and getting paid), allowed the patient's mother to die and leave him a vast inheritance and, it would seem, change his attitude toward her.

The system becomes even more complicated when patients do well and the therapist shares in the positive outcome and even extolls himself. If patients do poorly, however, the therapist explains that they are obviously resistant to therapy. It is a no-win situation for patients and a no-lose situation for therapists.

A basic problem in psychiatric treatment is that psychiatrists and parapsychiatrists (psychologists, psychiatric nurses, social workers, counselors, etc.) tend to want to make their patients "comfortable" in therapy. Individuals who are relatively comfortable would be ready and willing to attend their psychotherapeutic sessions regularly because they are stroking or pleasant experiences. It is my contention that individuals are willing to change only when they themselves are uncomfortable with aspects of their own psychological state or behavior. If the mental health professional makes individuals "comfortable," the necessary changes in the lifestyles or thinking processes will not occur. However, if individuals suffer from a mild degree of anxiety and/or depression, their willingness and readiness to change their behavior is more apparent. People will not change because another person wants them to. Change occurs only from within, and only if individuals want to change.

The desire to change must be from within. It is only accomplished if individuals suffer from anxiety and/or depression over their state of mental health. So "comfortable" psychological interviews are more often exercises in futility. It is futile to help patients achieve growth and psychological maturity, but it may be a positive for the therapist to entice patients to continue regular attendance at the prescribed "therapeutic hour."

It is also my contention that a psychiatric diagnosis is made only insofar as it relates to, and is in keeping with, the background of the individual making the diagnosis. I have had many patients referred to me by social workers who are desperate. They have diagnosed these individuals as having a "rich fantasy life." That corresponds to the social worker's frame of reference, but after completing objective psychiatric evaluations, it is obvious that these individuals are either hallucinating and/or delusional. The difference between hallucination and a rich fantasy life is based primarily on the background of the individual making the diagnosis, and the intensity or longevity of the relationship. The degree of training dictates the number of therapeutic options available, and also limits the number of diagnostic possibilities they entertain. The patient and the diagnosis must therefore fit into the practitioner's theoretical framework. Practitioners with more varied backgrounds are better able to fit the patient's needs rather than having the patient fit the practitioner's needs.

It should be noted that during the acute stages of psychiatric illness, it is far more important to help individuals reach a state of homeostasis (internal quiescence and solitude) in their psychological existence than to delve into their overall psychological states and backgrounds. During the acute states of an illness, it is and has always been my intention to relieve people from their overwhelming stresses and strains. Once homeostasis is achieved, however, moderate to significant amounts of stress are then placed on the individuals to help stimulate growth and learn to cope with reality. I feel this approach, although contrary to many commonly accepted psychiatric tenets, is far more useful and reliable. I apply these techniques to average individuals—not individuals with large bankbooks, but rather the overwhelming number of psychiatric patients whose resources are limited. These are people who cannot afford the luxury of illness, who must return to their families and jobs as soon as possible.

A significant number of my statements are controversial, if not heretical. I am not condemning every practioner of psychiatry, psychology, social work, or psychiatric nursing. I recognize that many of these individuals who utilize absurd techniques are open, honest, and knowledgeable individuals who genuinely believe in the system they are practicing. I am condemning the large group of mental health professionals who recognize the absurdity of their approaches and the ineffectiveness of their techniques. These individuals who continue to practice antiquated and often harmful techniques get our disdain. An honest mistake is quite different from a financial ruse.

The mental health field as a whole has been constricted by antiquated concepts and techniques, and has been caught up in the Peter Principle. This states that the longer an individual is involved in the same business and/or profession, the more likely the individual will reach the highest level of incompetence. This concept has caused the mainstay of psychiatry and psychology to remain in the dark ages for a considerable period of time.

Although this may sound somewhat harsh and critical, this

concept does have a concrete application. At this time, the majority of individuals who wield the most power in the areas of psychiatry and psychology, especially with reference to organizations and Boards, have been trained in the most antiquated techniques. Each and every practitioner of the mental health sciences must review and evaluate his own standards and approaches to mental health. In every profession, there must be growth and progress in order for that profession to be a productive part of society, and not a parasite. Even Freud, made many changes in his own theories. He was rather fluid in his approaches to patients, but the neo-Freudians have been less willing to accept new ideas and concepts. These people, not Freud himself, have stopped progress.

I must clearly acknowledge that at times, I have practiced unnecessary techniques and have made errors in judgment. However, I have since found other techniques to be more efficient. When I speak of the efficiency of techniques, I am referring to the vast majority of patients who are essentially Lithium deficient and are misdiagnosed and, consequently, mistreated. Hopefully, these misdiagnoses are honest, but I'm not always sure they are.

It is essential to my treatment approach that time be of the essence. I spent six years in general practice, and it is my contention that you must see that the results which are noticeable are a function of the techniques employed, not just the passing of time, or trial and error. When a person has an infected throat, the judicious use of an antibiotic causes a remission. However it is possible that another causative agent is responsible. If this is the case, additional visits and testing are appropriate. With psychiatric illness, this type of situation is also applicable, but only with the minority of patients. Most are preprogrammed to believe that additional visits are a necessary part of the overall treatment process.

In summary, when an individual is seen on a regular basis for an extended period of time for reasons of mental health, I am usually unclear as to whether or not the therapeutic approach in itself was useful or whether "tincture of time" and life circumstances were the cure-all.

The concept of "tincture of time" becomes even more complicated and ridiculous when you think about children and adolescents. They change their thinking on a regular basis, frequently and intensely, so if a therapist is seeing an adolescent as a patient and the patient changes significantly over a three- to five-year period, it is my contention that the therapist's role is, at best, questionable, if not totally unnecessary. Adolescents change their minds almost as frequently as their clothing. The mainstay of the treatment of adolescent turmoil is not time, but Lithium Carbonate.

I have found on repeated occasions that many patients who have attended their psychotherapeutic sessions regularly have continued to do so without noticeable results. I questioned these individuals at length and have found most of them to be Lithium deficient. Their fear of rejection, especially by an authority figure such as the therapist, more than overwhelmed and overrode their conscious knowledge of the absurdity of the system, so they continued to attend their sessions. Lithium sensitive individuals need to give "gifts" to other people. They may be material items or giving of oneself or one's time. To bring gifts to an authority individual is even more essential, necessary, and evident. These individuals, therefore, often subjugate their own needs for what they perceive to be the needs of the therapist. They also need to prevent rejection by their therapist, a recognized authority figure.

I realize again that these thoughts and ideas are contrary to current psychiatric thinking. Columbus, however, believed the world was round when everyone else thought it to be flat. Galileo refused to believe the earth was the center of the universe. The more I discuss these ideas with fellow psychiatrists, the more I am sure their own defensiveness colors their judgment. They seem to refuse to recognize the overall simplicity of treatment for the majority of patients. They appear rooted in tradition, irrespective of its ineffectiveness.

I recognize that my approach to the treatment of individuals whom I categorize diagnostically as minimally Lithium deficient is contrary to commonly accepted psychiatric theory; however, I have found commonly accepted psychiatric theory basically useless in the treatment of patients. Most individuals in today's society are Lithium deficient. Their deficiency presents itself in a wide range of symptoms which have created many diagnostic possibilities. The common denominator remains a basic deficiency in the mineral Lithium. Judicious application of minimal amounts of Lithium has resulted in the "cure" of many individuals who otherwise would have been subjected to a lifetime of emotional stress and illness. Fortunately, more and more accounts are being reported by many different people about the widespread application of Lithium Carbonate in an ever increasing number of emotional disorders.

The best system is the one that works. The best treatment is the one which offers the maximum relief to the maximum number of people. The best treatment is the one that takes the least amount of time and is the least expensive. I contend that the system I use is, without question, the most efficient and inexpensive, as well as being readily accepted by the vast majority of patients to whom it is offered.

Furthermore, it is my experience that an overwhelming number of individuals suffer from a number of variations on the basic theme of Lithium deficiency. If they are offered Lithium Carbonate therapy early in the course of their emotional difficulty, most will respond rapidly and be able to rejoin their social environment and work force with very little difficulty and with minimal interruption in their basic lifestyles.

Chapter Seven

NOT TO USE IT—IS TO ABUSE IT

It is my understanding and experience that throughout the practice of psychiatry the best way to get rapid and successful treatment of a very severe disorder is for the patient simply not to have had it. The vast majority of schizophrenic patients being treated in this country aren't schizophrenic. They are neither schizophrenic, nor are they being treated.

There is a myth called "The Rule of Thirds" which states that one-third of the individuals who are diagnosed as schizophrenic will never get well, one-third will have recurrent, acute attacks with intermittent quiescent periods, and one-third will get well and never have another episode. The latter two-thirds never had the illness to begin with, so if a psychiatrist wants to have a high statistical success rate in the treatment of schizophrenia, he can readily do so with almost any approach or new medication because most individuals, at least the two-thirds who were diagnosed as having schizophrenia, did not have it at all.

The vast majority of psychiatric patients with whom I have dealt have first been treated elsewhere. Their treatment usually falls into the "tincture of time" category. Given enough time, most psychiatric illnesses will respond, irrespective of the therapist or the therapeutic approach. In addition individuals at risk often do better at home than they do in a psychiatric setting because the home is generally a more pleasant atmosphere than a psychiatric hospital.

The usual method used for psychiatric treatment and care is whatever method is in vogue at the institute where the individual is being treated. If the patient is an outpatient in a private psychiatric practice, the most prevalent, but not always the most appropriate, system used is the one, and usually the only one the therapist knows. To reiterate, the average therapist is one who is primarily trained in one modality of treatment and may or may not have superficial knowledge or experience in another one or two areas. There are many approved types of therapeutic interventions and dozens upon dozens of major accepted techniques within the United States. The criteria which establish the efficacy of one system over another is not the patients' needs or requirements, but in the vast majority of cases, the place they happened to have gone to or the person they happened to be seeing for help.

To further complicate the situation, individuals who each know a particular modality or come from a particular training program are both honor bound and knowledge bound in that one area, so that one system must be used. They each feel their system is the ultimate system, appropriate for every and any type of problem that exists in the spectrum of psychiatric illnesses. Their system is the ultimate system; their truth, the ultimate truth. However, each different school of thought in the mental health field proposes to have within its province the ultimate truth.

Another complication is the age of the therapist which has a great deal to do with the extensiveness of training in the different aspects of a particular therapeutic approach. At different points in years past, different aspects of the same approach have been stressed in the learning institutions and hospitals where therapists would take up residency or training. So not only will average therapists obtain the overwhelming preponderance of training in one modality of treatment, but most likely only the one aspect which is in vogue will be stressed. As in any field, progress is constantly being made in the mental health field. It is most important for all professionals in the field to keep up with new discoveries and treatments. If therapists feel they already know the "ultimate truth," they are limiting themselves to the archaic principles usually learned many years before. Keeping the mind open to new techniques and schools of thought seems to be limited when observing the attitudes of most mental health professionals.

Another major factor that complicates the situation is the individual's pocketbook. The number of treatment sessions per week, the overall length of treatment, and the additional ancillary techniques are usually a direct function of the ability to pay. Patients with significant financial backing are "appropriate" for long-term, analytic therapy. This may be anywhere from as little as one session per week to as many as five sessions per week, for an unknown or indeterminate period of time. The availability of insurance programs, as well as actual dollars must be considered in order to fully understand this system. It is rare for a mental health professional to make a diagnosis that isn't maximally profitable. The diagnostic nomenclature, the treatment procedure, and the overall approach is a direct function of the length of the therapist's waiting list and the need for a patient hour to be filled. The less training an individual has, the less accurate the diagnosis.

In the mental health profession, we deal not only with psychiatrists, but with a host of semi-professional professionals. Some of my best friends are "psychologists." I am not casting disparaging remarks toward psychologists in general. I am speaking specifically about the limited number of diagnostic categories a clinical psychologist can treat. To continue along the lines that no one makes a diagnosis that isn't maximally profitable, psychologists, especially those not working closely with psychiatrists, frequently miss the manic-depressive diagnosis. Even though clinically they can pick it up in testing, they often overlook it when dealing with patients, as in a treatment-oriented setting.

The situation deteriorates markedly when we enter the area of the diagnostic capabilities and qualifications of psychiatric social workers, people who have degrees in counseling, and/or psychiatric nursing.

Over the years, I have had referrals of some of the most psychiatrically ill patients who had been seeing lower echelon psychotherapists. These individuals were flagrantly psychotic, hallucinatory, delusional, and, at times presented a clear and present danger to themselves and/or others. Lower echelon psychotherapists may be unwilling to see psychopathology because if a person is severely ill he will go to a psychiatrist for more effective treatment and be lost to the lower echelon professionals. Perhaps the mistakes they make come from good intent but poor judgement. The answers are for the most part unclear. I must say I have also seen and worked with some extremely competent individuals in all levels of the mental health professions. Additionally, I have dealt with many shortsighted, ill–mannered, educationally limited and backward psychiatrists. The main reason I stress competence versus less than competent is because the overall treatment modality of manic-depressive illness, (or the seventy-five percent of the population that is Lithium sensitive to some degree or other), could be managed simply and readily by moderately trained individuals who have proper supervision. By law, a physician must manage the prescribing of Lithium Carbonate, however anyone who truly understands the varied degrees of the personality characteristics of Lithium sensitive individuals can educate and treat Lithium deficient individuals. It has been my experience that the team approach is the most efficient and the least expensive.

It is my aim and hope that the general practitioner of medicine and surgery will recognize the simplicity of using Lithium Carbonate in low level doses. The diagnosis of Lithium deficiency is relatively easy if the doctor allows himself to make it. If he were to prescribe two-to-three capsules per day after making this diagnosis, he could actually treat most individuals who require Lithium Carbonate without the services of a psychiatrist. I would recommend that if people need more than three or four capsules of Lithium per day, they should be seen in conjunction with a psychiatric consultant.

General practitioners are the mainstay of medicine. They are the front lines of almost all medical, surgical, gynecological, pediatric, and ophthalmological illnesses, as well as many others. Therefore why should they not be the mainstay, the forefront, for treatment of the varying degrees of manic-depressive illness. Most general practitioners will attest to a large percentage, if not the majority, of their overall practice as psychotherapeutic anyway. Why then not give them some extra tools with which to work. I see my role as an educator, not only to teach patients but to teach other professionals equally, if not more so.

Lithium Carbonate is well known as the treatment of choice for manic-depressive illness. Let me stress again that Lithium Carbonate is a mineral that naturally occurrs in nature. It comes from the ground, like potassium and sodium. Most of our mineral supply of Lithium has been taken away from us because of pollution of the water supply and filtration. During filtration most of the Lithium is extracted, and therefore what is left is insufficient for our needs. We can get some natural Lithium from kelp, but this is only minimal.

Lithium Carbonate is a salt. In ancient Rome, a great orator named Cato ended every speech with the statement, "And Carthage must be destroyed." Irrespective of his thoughts, ideas, subject matter, or presentations, every time he met a friend or yelled at an enemy, and every time he greeted someone or left them, he ended with the statement, "And Carthage must be destroyed." In time he was able to literally program an entire nation to believe that "Carthage must be destroyed." Rome then initiated the third and last of the Punic Wars. So every time I see a patient who must go on Lithium, a patient who is on Lithium, or enter into a discussion with anyone about Lithium Carbonate, I begin and end with the phrase, "And Lithium is a salt; it is a naturally occurring mineral in nature." I continue to stress this so stringently because most people, no matter how many times I tell them, say "Yes, but I don't want to take drugs for the rest of my life." Even though Lithium is sold in drugstores, it is not what would classically be described as a drug. It is a naturally occurring mineral in nature. I stress this over and over, to alert the adult population and also the adolescent population and the naturalists.

Lithium is to manic-depressive illness as insulin is to diabetes. Diabetics need an external supply of insulin in order to function. Some early diabetics, or prediabetics, just need dietary control. By comparison, individuals in the early stages of manic-depressive illness might do well with kelp and/or merely one capsule of Lithium per day. There are moderate diabetics who need one injection of insulin per day. This injection may be comprised of more than one type of insulin. A moderately Lithium deficient individual may need two or perhaps three capsules of Lithium per day to help maintain an emotional state of relatively calm consistency. As we consider more severe types of diabetes, we must consider individuals who need more than one dose of insulin per day and perhaps more than one type of insulin in each dose. At this end of the scale, we refer to the most severe. We also refer to the most severe individuals who suffer from the many variations of manic-depressive illness. These individuals usually require hospitalization, at least at the onset, and the use of perhaps six or eight capsules of Lithium Carbonate per day.

To continue this analogy: a diabetic needs sixty units of insulin; those sixty units are established with a relationship to a specific diet, and the patient is expected to follow the diet and take the appropriate amount of insulin. Lithium deficient individuals who require three or four capsules per day are expected to take three or four capsules per day on a regular basis. They are also expected to maintain the minimum dietary restrictions necessary to have Lithium function at its optimum capacity. These restrictions are primarily the relatively consistent daily intake of fluids and sodium chloride.

When a diabetic first comes to a doctor for a consultation and it is established that insulin is necessary, how much insulin is necessary cannot be determined at that first meeting.

In my general practice, when a patient would ask, "How much insulin do I need?" I would respond, "I don't know." The patient would often look at me quizzically. I would then explain that when I say I don't know, I am saying we have a certain end point we want to accomplish. That end point is a certain level of blood sugar. How much and how often insulin will be needed can only be established after observing the changes in blood sugar levels when additional insulin is added to the body's system. Dietary needs also have to be established according to the individual's physical needs. A bookkeeper requires less calories than a lumberjack to maintain daily work activities. When we consider a patient's diet, we must consider optimum weight, the daily expenditure of exercise calories, and the level of rest. We have to account not only for the patient's work but for the total activity level.

We also have to take into account the relative degree of cooperation we can expect from the individual on an ongoing basis. This is absolutely crucial. It has been my experience that regardless of the illness, the longer they feel well, the less likely they are to cooperate in long-term treatment and follow-up.

The longer average diabetics are on insulin, the more likely they are to begin varying their diet and "cheating," the more likely they take their insulin requirements more lightly and forget how ill they were prior to the onset of its use.

Who is the average patient requiring Lithium Carbonate therapy? The criteria for treatment with the use of Lithium Carbonate are yet to be established and vary greatly from individual to individual. My earlier references to a treatment program being established primarily by the experience of the therapist and by the desires and finances of the patient hold true. My practice was originally geared to a lower middle-class population with a finite amount of money available to them for "psychotherapy." Therefore, I began using a number of behavioral and hypnotic techniques to help effect rapid changes in symptomatology and personality. My practice flourished. I began seeing large numbers of people for very short periods of time. Essentially, I was once again a general practitioner, but this time I was a general practitioner in the field of psychiatry.

I began to see some patients with symptoms compatible with the diagnosis of manic-depressive psychosis who had to be hospitalized and treated in an intensive manner in a controlled setting. I was also able to see members of these people's families and discovered that they had similar personality characteristics, but usually to a lesser degree. They were worried about these similarities that were interfering with their ability to live life to the fullest. These characteristics were deterring them from being able to fully reach their potential. They spent much of their time feeling like an eight-cylinder engine operating on only five or six cylinders, sometimes on seven, and sometimes only on two or three, but never really totally shut off or fully powered.

I began to establish a list of "criteria of a manic," or what

I call a Manic's Manual. I established a list of characteristics I most often saw in individuals who required Lithium Carbonate therapy. I was able to establish very quickly that the individuals who suffered from this deficiency, but to a milder degree, had characteristics that were very similar, if not the same, as those who required large amounts of Lithium Carbonate. Manic-depressive illness is a matter of degree, not kind.

The degree to which an individual suffers more or less dictates the amount of Lithium Carbonate therapy necessary. Most people I see require only minimal amounts. One capsule, two to three times daily seems to be the amount of Lithium required by the vast majority of individuals. At the present time, I am treating many doctors, lawyers, dentists, psychologists, professors, type A successful businessmen and type A not so successful businessmen with Lithium Carbonate therapy. The success of a businessman is not how much money or how many positions he has held, but rather the amount of money he keeps or the positions he can stay with and in which he can flourish.

I do not wish to pretend for a moment that Lithium Carbonate is the answer to all the world's problems, the ultimate panacea. Lithium Carbonate essentially allows people to maintain cognitive or thinking control over intense, emotional responses. It allows them to put their brain in gear and then to put their mouth in motion. It provides an atmosphere of quiescent consistency so individuals can take stock of themselves and how they respond to their internal and external environments. Most patients report that Lithium gives them a sense of inner calm and peacefulness. It makes them feel as if their engine is no longer running overtime. They describe a sense of peace and unburdening. They feel they can keep up with their thoughts, even to the point of controlling them and they do not feel as if they are spinning in twelve directions at once and running from one place to another. They no longer have to spend a great deal of energy to accomplish a minimum amount of work. There is a new-found ability to fall asleep; their minds are no longer spinning to the point where they cannot fully rest or sleep; they have a sense of awareness of themselves and their environment, and an ability to recognize the needs of others around them.

Many Lithium deficient individuals have said how wonderful it is to be able to truly see, smell, and even taste, and how their hearing and skin sensitivity changes when their Lithium deficiency is eliminated.

When individuals who are Lithium deficient are given appropriate replacements of Lithium Carbonate, they are able to turn outside of themselves and see the world as it is. Their rose-colored glasses are shattered. Their gray glasses of depression are equally shattered. They can appreciate nature and the wonders around them. Many of the things they see as if for the first time can be a significant discovery: colors are brighter, sounds are sharper, and smells are crisper.

Most individuals who have intense "ups and downs" describe with a sense of grief and despair that "my highs are gone." They describe a sense of loss over their manic "grandiose, indestructible periods." However, they quickly counter with the fact that "I understand why I won't have them anymore, and I understand why they must go so I can finally control these overwhelmingly painful lows. However, I still miss them." A manic loses his highs within a matter of days to weeks, yet it often takes months to control the lows. Most individuals suffering from this manic-depressive illness or Lithium deficiency take six to nine, sometimes even twelve months for the lows to become fully under control. During that time, the lows are usually less intense, less frequent, and of shorter duration. It is often necessary to use a more conventional antidepressant during this time. The use of an antidepressant is a standard medical and psychiatric approach. The price paid for the loss of the highs is moderate to nonexistent compared to the price an individual paid continuously for the lows.

Lithium deficient individuals must be instructed about lag time. There is often a four to six week period between the time people may appear outwardly to be coming out of their depression vs how they still feel internally. As we have stressed, Lithium deficient individuals most of all fear rejection, after which comes isolation or loneliness, followed by lying, being lied to, or being patronized. It is important to let Lithium deficient people know that this lag time exists. Their gaits become bouncier, their faces more animated, their movements more purposeful, their overall demeanors more at ease.

It is much easier to explain something before it happens, rather than trying to spend hours regaining their confidence if they feel you have either held something back from them or are patronizing them.

Lithium sensitive individuals, after being on Lithium Carbonate therapy for several months, are then patients who are suitable for psychotherapy. When individuals who suffer deeply from diabetes enter a hospital in a state of coma or semicoma, they are placed on insulin intravenously to help them deal with the acute phase of the illness. Only after they fully regain control over their body functions are they placed on dietary control and self-administration of insulin. The parallel remains: Only after Lithium deficient individuals are on Lithium therapy for a significant period of time are they amenable to using routine techniques of psychotherapy. These routine techniques are then usable because the individuals have obtained a consistent mood level. Without a consistent mood level, no technique can stand the test of time.

Individuals do not become androids or robots by taking Lithium, but rather the extreme peaks and the extreme valleys of mood swings are brought under control. These individuals still react with a full range of emotions; the overwhelming intensity of the reactions and their unpredictableness are altered. Individuals who are on Lithium, although having somewhat blunted or lowered energy levels compared to the highs of the non-Lithium state, will accomplish more. Instead of having many thoughts racing or many projects underway at one time, the numbers of thoughts and projects will be significantly decreased. The outcome will be significantly more productive and rewarding. A far greater amount will be accomplished, and accomplished well.

Blood tests are necessary to the overall management of Lithium sensitive individuals who go on Lithium therapy. Just as the blood sugar has to be measured for diabetics, the Lithium level has to be measured for Lithium sensitive patients. The dangers of Lithium Carbonate have been overdramatized and because of this, too many blood tests have been drawn. It has been my clinical experience that the vast majority of individuals who are Lithium sensitive, after the first few months of weekly, or perhaps twice weekly tests, need blood tests only every two to three months. If there are any significant personality changes, I recommend that all Lithium therapy patients get a Lithium blood level drawn at that time to insure safety and also to insure proper patient acceptance.

I do not stress blood tests. I stress the need for patients to take their mineral supplement medication regularly. I reinforce the fact that they are in control of their destiny now, and that whether or not they take their Lithium is up to them. They do not need a policeman watching over them.

I fully elaborate all the warnings that are appropriate for Lithium takers to know. I also provide written material on the use of Lithium Carbonate and any side effects which might or might not occur. Side effects are common with all medications, from birth control pills to digitalis. All medications can cause severe harm to the human body if taken improperly and if managed improperly. It has been my clinical experience that the complications of Lithium Carbonate therapy are extremely minimal. There are, however, several patients who have had what is called an idiosyncratic reaction. This means their reaction is out of keeping with the norm, it is unusual and unexpected. These are individuals who could not tolerate even one capsule per day. Fortunately, Lithium also comes in a liquid form which makes the dosage more adjustable. With the judicious use of increased amounts of Lithium over time, I was eventually able to augment their Lithium deficiency to the point of personal control.

When Lithium sensitive individuals first start on Lithium therapy, the lengthening of their fuses, the loss of impatience, the ability to go outside and play with the kids, and the ability to want to be part of the family are quickly noticed.

Another major aspect of the treatment program of manic-depressive individuals is the inclusion of the family for educational as well as explanatory reasons. Family members need a place to vent their anger, frustration, hostility, and their own rebound emotional reactions. I have found that family involvement and education has been valuable in the overall approach to the comprehensive treatment of this illness.

Any long-term or chronic problem affects the individuals at risk and also the individuals around them. The family also suffers, and their needs must be acknowledged and answered, too. Their role in treatment is almost as important as that of the Lithium sensitive individuals themselves. Part of the outcome of "getting well" is the maintenance of the family structure as a whole. We are all social beings. We live in a two-by-two world. Preservation of this concept should, therefore, go hand in hand with the treatment of a Lithium deficient husband or wife.

Chapter Eight

NATURE vs. NURTURE

There has been a longstanding, ongoing controversy within the fields of psychology and psychiatry about whether genetics or environment is the most prominent factor in the formation of an individual's personality style. My clinical experience has shown me overwhelmingly and without a doubt that environmental conditions are at best secondary to genetic endowment. I therefore believe that an individual's genetic endowment is without question the most crucial factor in personality formation and personality style. I am intentionally repeating myself for emphasis. This basic tenet is crucial for the understanding of Lithium sensitive individuals and their treatment.

We will now develop the concept of "personality quotient." This is applied to an individual who is born with a particular personality characteristic, to a greater or lesser degree. Whether or not this characteristic is relatively dominant or is less readily observed may well be the function of environment and upbringing.

The most powerful and overriding emotion and characteristic of average Lithium deficient individuals is the combination of (a) the fear of rejection, and (b) the fear of being alone (the fear of being alone defined as one which occurs when an individual is left to one's own devices, not by choice, usually without a specific time frame about when the other individual might return).

Lithium sensitive individuals who have, for example, a genetic sensitivity to rejection of sixty to seventy percent may be able to hold the feelings of rejection somewhat at bay if they live in a very warm, positive, loving, caring environment. However, should they live in a very caustic, attacking, critical, and negative environment, they may develop this characteristic to a far greater degree. Lithium sensitive "feelers" are extremely sensitive to environmental circumstances and input. This is such a great influence on their mental health that they react to any real or perceived fluctuation from absolute, positive input in a most dramatic and overwhelming way.

The influence of the environment in the development of an individual's personality is widely accepted. However, irrespective of the degree of positive input to which "feelers" are exposed, when people have a high degree of inferiority, insecurity, and self-worthlessness as a specific genetic endowment, they will nonetheless maintain these characteristics. These genetic characteristics themselves will have a dramatic influence on the "feelers'" state of mental health. People whose mental health is relatively dependent on the external environment remain labile and easily influenced. Depending on the specific genetic endowment, individuals are programmed through their own DNA and RNA (the genetic transmitters) in a predetermined manner. DNA and RNA are the substances which allow procreation to continue. It would appear more natural than not for the same characteristics to be passed from generation to generation. Additionally, you must keep in mind that within a given family structure, similar genetic patterns exist. Family members share similar sensitivities and reaction patterns. The phrase "like father, like son" is quite appropriate, however, even if the son never had contact with the father (environmental exposure), the phrase would still be true.

You may be getting a sense of predestination. However, there remains the freedom of choice to be able to overcome and counteract this set of conditions. The ability to counteract and, consequently, to maintain control over these characteristics is directly related to the individual's willingness and ability to take mineral supplements. Individuals who are willing to accept their frailties and genetic predetermination can become independent of them with the use of Lithium Carbonate. Lithium Carbonate, once again, is a natural mineral substance which allows an individual to obtain and then maintain cognitive or brain control over his emotional unbridled heart needs and drives. When individuals accept the fact that they are, in essence, Lithium deficient, they then place themselves in a situation where perhaps for the first time in their lives, they become free thinking, self-controlled, and self-actuated. Perhaps for the first time in their lives, they will be able to maintain cognitive control over their visceral needs.

When Lithium deficient individuals take Lithium and their moods become consistent, they begin to trust themselves. They begin to be able to make plans, confident that they can carry them out. Other people begin to know how to relate to them and stop "walking on egg shells." Their lifestyles change, become more consistent, more positive, and more friendly.

Individuals must have consistency in order to function. Lithium deficient individuals may function adequately or inadequately, but without consistency, they cannot function at all in any manner which would create a positive relationship with others or even within themselves.

The crucial point I want every Lithium deficient "feeler" to understand and feel is : WITHOUT CONSISTENCY, THERE CAN BE NO TRUST. WITHOUT TRUST, THERE CAN BE NO RESPECT. WITHOUT RESPECT, THERE CAN BE NO LOVE. AND, WHAT IS LIFE IF THERE IS NO LOVE? Without Lithium the majority of people are inconsistent. It's the only place to start the desired cycle and obtain consistency.

Chapter Nine

JUST ANSWER THESE QUESTIONS— YOU'LL KNOW!

I have presented a great deal of information about Lithium deficient individuals. I have stated that I believe that at least three quarters of the population could benefit to some degree from adjunctive Lithium mineral replacement. It is now time for you to assess whether or not Lithium Carbonate therapy would be helpful to you.

Take a sheet of paper and number it from 1 to 75. Answer the following questions as honestly and as objectively as you can. Do not answer them the way you would want others to see you, but rather as you really feel you are. Do not answer them as you would like to be, but as you actually perceive yourself to be at this moment in time.

Each answer must be a simple yes or no. Answer the questions immediately and decisively. Put your immediate "gut" response on the proper line. Do not stop to think deeply, or to question your response. Just write it down.

Remember, the answer you give is to be based on how you actually feel on the inside, not how you appear or would like to appear on the outside. You are the only one who needs to see the answers, so be as open and honest as you can. The results may surprise you.

	YES	*NO*
1. Do you have a significant or overwhelming concern about rejection? or criticism?	____	____
2. Is peer acceptance noticeably important to you?	____	____
3. Do you have an underlying sense of insecurity and lack of self-confidence, despite		

the accomplishments you may have achieved in life? ____ ____

4. Do you feel you do not deserve the accomplishments you have attained? ____ ____
5. Do you request the assistance of others to do relatively simple tasks which you actually can do on your own? Do you have difficulty dealing with simple problems without other people's opinions? ____ ____
6. Do you tend to overly criticize others for lying, or have you found yourself overreacting or being unusually sensitive to "white lies?" ____ ____
7. Have you found yourself reacting very strongly when someone disagrees with your opinion? ____ ____
8. If someone does not like something you do or own, do you become overly upset or overreact? Are your possessions or opinions actually you? ____ ____
9. Are you a hopeless romantic? ____ ____
10. Are you a Don Quixote? Do you tend to see the world and the people in it the way you want to see them, rather than as they are? Do you wear rose-colored glasses? ____ ____
11. Do you suffer from the Dulcinea Syndrome? ____ ____
12. Do you attempt to create the reverse Dulcinea Syndrome? Are you an individual who attempts to create an image of yourself so that others can view you the way you would like them to? This image is usually created despite the pain endured or the consequences at hand. ____ ____
13. Are you unusually possessive of others by giving too much? Before you answer this question, think of this: The most possessive people are the ones who give, and give, and yet give some more. By giving to another

in an inordinate amount, the individual who is the receiver is placed in a position of being obligated to the giver. This obligation is initially quite subtle. However, as time goes on, the receiver feels more and more uncomfortable being given to. The giver cannot "understand" why the receiver rebels. To restate the question: Do you tend to be unusually possessive via the modality of giving? ____ ____

14. Can you imagine and/or develop projects in your own mind? Do you tend to be unusually creative? ____ ____
15. Despite your ability to create, do you tend to lack follow-through? Do you seem to simultaneously have a number of unfinished projects? ____ ____
16. Are you an individual who tends to think in pictures and images, rather than in words or lists? ____ ____
17. Are you known to be unusually verbal, possibly to the point of having logorrhea (diarrhea of the mouth)? ____ ____
18. Do you have, or have you had in the past, a tendency to be overtrusting and undercritical, especially of friends and loved ones? ____ ____
19. Do you tend to believe that there is a supernatural force, be it called God or something else? ____ ____
20. Do you maintain exceptional personal responsibility in the area of your professional life? ____ ____
21. Do you tend to assume added and, at times, unnecessary responsibilities at work? Are you the all-time volunteer? ____ ____
22. Is your basic orientation one in which you want to be told what has to be done, and

not how it has to be done? Are you an individual who finds that if you are told why something has to be done and how to do it, that you become bored and lose concentration on the subject? ____ ____

23. After you have created an idea, do you tend to find another individual and set it up for him to carry out for you? ____ ____
24. Do you either consciously or subconsciously allow another individual to volunteer to help, so not to have to go it alone? ____ ____
25. Are you known to be relatively impulsive? This particular question needs amplification. Individuals who suffer from a circular mood disorder, Lithium deficiency, have difficulty making long-range plans. They are often unsure on Monday what their mood will be like on Friday. Often, unfortunately, they are unsure on Monday morning what their mood will be like on Monday afternoon. Therefore, when a Lithium deficient individual recognizes that they have the energy level or drive to carry out a plan, which itself may have been waiting for quite some time, they "impulsively" carry out the plan. This impulsive action is rather a lack of ability to trust one's moods and to make long-range plans, rather than being classically impulsive. Do you, therefore, tend to be impulsive? ____ ____
26. Are you an individual who is unusually subject to peer pressure? For example, if you want to go to the movies but the rest of the group wants to go to a club, will you go with them rather than not be part of the "crowd?" ____ ____
27. Are you an individual who has difficulty learning from history? Do you tend to make

the same mistakes over and over, especially in the area of interpersonal relationships? ___ ___

28. Do you tend to daydream a great deal? ___ ___
29. Have you been known to be so focused on a television show or a novel that someone had to call your name more than once to get your attention? ___ ___
30. Have you had highway hypnosis? Have you driven down the road and suddenly realized that you had covered several miles, stayed in your lane, maintained your speed, and yet consciously were unaware of your actual driving? ___ ___
31. Are you known to be impatient, or do you feel impatient internally even if you don't always show it externally? ___ ___
32. Are you known to have a tendency to overreact? This particular question is one of the most important guidelines in establishing the diagnosis of Lithium deficiency. An individual overreacts without "thinking." Their "feeling" nature tends to predominate. ___ ___
33. Are you known to be a person whose moods change frequently and unpredictably? When answering this question, be sure that you are answering it in a manner which represents the way you actually feel, not necessarily the way you may appear to others. A large number of Lithium deficient individuals have mastered the ability of "hiding" their feelings, yet they internally experience the same degree of emotional pain as individuals who wear their hearts on their sleeves. ___ ___
34. The eggshell test: Do people around you feel that when dealing with you they are

walking on eggshells? Are you the type of individual who makes others have to think and carefully plan when they can approach you with certain questions so that they can "hit you in the right mood"? ____ ____

35. Are you an individual who can be set up to react rather than to think through a situation and then act? Think for a moment about this question. Think back for a moment and review situations of an emotional nature which may be repetitive. For example, if you have a child who leaves a bicycle in the driveway, is your reaction predictable? Do you tend to yell, get upset, and yet actually do nothing about it? Guess what? You've been set up. ____ ____
36. Are you an individual who tends to answer questions rapidly and without thinking? Do you tend to put your mouth in motion before you put your brain in gear? ____ ____
37. Do you tend to answer questions even before they have been fully posed to you? ____ ____
38. Do you recognize that you may be known as an individual who has an attitude of "would you mind not talking while I'm interrupting"? Individuals who are Lithium deficient tend to interrupt others. They do not intend to be rude or inconsiderate, yet they interrupt frequently. The reason for this is that as a thought pops into the mind of a here and now person, the thought is immediately expressed. Do you find that you interrupt others while they are speaking? ____ ____
39. Are you an individual who needs comments made about a job you have done? Would you feel uncomfortable and at times some-

what anxious unless your boss says "nice job"? ____ ____

40. Do you embarrass easily? ____ ____
41. Do you need to be stroked (complimented) but not massaged (complimented) too intensely or too frequently? ____ ____
42. This two-part question is another one of primary importance. Do you tend to anticipate the behavior of others? Do you frequently become disappointed when others have not acted in the manner in which you wanted them to, even though you didn't verbally tell them what your expectations of their behavior were? ____ ____
43. Do you find that you are indecisive on minor issues? For example, do you find that it would be easier to decide whether or not you need a new home than to decide what color the house should be? Do you find it easier to decide whether or not you want to go on vacation than where to go? ____ ____
44. Are you known as a "kissy, huggy, touchy" type? For example, do you like to hold hands and have yours held equally as much? ____ ____
45. When in a store, do you tend to pick something up to examine it, rather than just look at it? ____ ____
46. Do you tend to have difficulty scheduling your time? Do you tend to take on more than any human being can possibly handle? ____ ____
47. Are you the type of individual who always thinks there will be enough time later on in the day to get the job done? ____ ____
48. Are you the type of individual who has a good deal of difficulty working alone? This

is a repetitive but essential theme in the life of a Lithium sensitive individual. ____ ____

49. Another most important question: Are you an individual who has difficulty accepting unsolicited criticism? ____ ____
50. Are you an individual who feels crushed if you receive negative criticism from a peer or superior when completing a task? Do you feel that this negative criticism is an overall global attack on you as a person rather than on the particular job or situation at hand? ____ ____
51. Are you a bit of a showman? Do you tend to stand out in a crowd? Do you actually enjoy it? ____ ____
52. Do you tend to have frequent migraine headaches, colitis problems, general chronic fatigue, etc.? Do you tend to go to the doctor at an unusual rate? ____ ____
53. This next question requires additional explanation. As we have mentioned, Lithium deficient individuals are known as visceral, emotional individuals. They are individuals who tend to speak through their viscera, that is to say it is easier and more within keeping of the "feeling" computer to develop psychosomatic illnesses. For example, Lithium sensitive individuals who are uncomfortable in a particular situation and feel as if "something is eating at them" may develop an ulcer rather than verbally confront the situation. Lithium sensitive individuals do not do well with direct confrontation because they are overwhelmed by the fact that their rejection button may be pushed. The question here is: Do you tend to have a noticeable number of "psychosomatic complaints"? ____ ____

54. Are you the type of individual to whom the answer "no" becomes an overwhelming and dramatic situation? Does "no" devastate and immobilize you? ____ ____
55. Are you known as an individual who is inconsistent, or do you feel inconsistent within yourself? ____ ____
56. Are you known as an individual whose child rearing techniques vary greatly from day to day, or even from minute to minute? ____ ____
57. Do you realize that you may well be the type of individual who is in need of an external support system for personal security? Do you need to hear positives from others to feel secure within yourself? ____ ____
58. Are you the type of individual who needs the input of others in order to appreciate yourself or a job that you have done? ____ ____
59. Do you tend to place more value and importance on what others say about you or about what you are doing than to trust in your own judgment? ____ ____
60. Do you notice that you tend to have significant variations in your memory? Do you notice that at times you may have forgotten where you left your car keys, yet you can remember clearly things you did years ago? ____ ____
61. When in a restaurant, do you have difficulty choosing what to eat? Do you find large menus more of a problem to deal with than ones with but a few choices? Would you prefer that others ordered first? ____ ____
62. Are you the type of individual who has significant concern about being in control? Remember, the ultimate of control is knowing when to be in control of giving it up. This is an unusually important and painful lesson to learn. ____ ____

63. Do you tend to create within your own mind's eye an image of another individual with whom you are dealing? ____ ____
64. Do you have difficulty just being yourself? ____ ____
65. Do you tend to react intensely if you feel you are being patronized or spoken down to? ____ ____
66. Do you tend to handle problems like an ostrich, burying your head in the sand to avoid making decisions? Do you prefer that others make these decisions for you? ____ ____
67. Do you feel at times as if your mind is racing away from you, and you can't keep up with your thoughts? ____ ____
68. Do you find that there are noticeable time periods when your thoughts and movements are running very slowly, and that you really don't give a damn? ____ ____
69. Have you ever noticed that you had difficulty separating "honesty" from "volunteering unnecessary information"?
    Have you noticed or been told that you are the type of individual whose need to volunteer unnecessary information is to their own detriment? ____ ____
70. Have you noticed that you are the type of individual who tends to repeat life's patterns, or have you noticed that you are the type of individual who does not necessarily learn well from experience? ____ ____
71. Are you an individual whose time frame reference is rooted in the present? Do you tend to think about the present much more frequently than the past or future? ____ ____
72. Are you the type of individual who is much more comfortable giving than receiving? Is it easier for you to buy a gift for another

than to allow that person to buy a gift for you? ____ ____

73. When depressed, do you notice that your behavior is to act as if you are attempting to "fill a void"? Do you notice that when you are down in the dumps, you feel as if something is missing, and yet cannot get a handle on what the something is? ____ ____

74. Are you known as an individual who has significant difficulty being alone? Are you the type of individual who, if you do not choose to be alone but are in a situation where you have to be alone, acts like a cat on a hot tin roof? ____ ____

This next three part question is reserved for our female readers.

75. Are you predisposed to unusual emotional difficulties and /or psychosomatic difficulties (headaches, etc.) during the premenstrual phase of the menstrual cycle? Do these difficulties continue during menses? Did you have a postpartum depression? ____ ____

It is now time to assess your score. The interesting part of evaluating a score of this nature is that there is no right or wrong result. You must be willing, however, to accept the bottom line conclusion. Look over your answer sheet. Count the number of "yes" responses. I am sure that the majority of readers who answer these questions honestly will find an overwhelming number of "yes" responses. Of course, certain questions are more heavily weighted than others. The positive responses to certain questions alone are almost diagnostic of a need for Lithium Carbonate. Specifically, questions referring to moodiness and unexplained inconsistency are of primary importance. Additionally, the questions about females suffering from significant mood changes around their menses or childbirth are equally important. Questions dealing with the concept of rejection and concerns of isolation or loneliness are likewise of equal importance.

If the majority of answers to the aforementioned questions are affirmative, you are suffering from a Lithium deficiency.

I recognize that I have stated this fact in a rather dogmatic fashion. Many of my patients have been unwilling to accept my appraisal on face value. This unwillingness is consistent with the medical community in which they live.

To help obviate this problem, we have established a "Lithium Club." This club (renamed the "Mood Menders" by one of its members) is comprised solely of individuals who are taking Lithium. The purpose is to allow individuals to ask any and all questions of people who are taking Lithium. In addition, members of families of Lithium responders and Lithium non-responders attend. There are individuals who have responded dramatically to Lithium Carbonate, as well as individuals whose responses have been minimal and/or slow. Entire families are welcome. All questions are answered by other members and during these meetings, there is no professional input and no intimidation by having professionals present. The Lithium Club has given us an additional and useful framework. Obviously, I am enthusiastically pro-Lithium. Often, individuals are overwhelmed by the concept of needing to take Lithium. By discussing their concerns with members of the Lithium Club, most of their concerns are rapidly dissipated. Group acceptance and the knowledge that others are taking Lithium has a very quieting effect on the psyches of Lithium deficient individuals.

We have mentioned frequently that Lithium sensitive individuals are verbal and tend to want to be part of a group. They are uncomfortable when they feel isolated and alone.

With the creation of the Lithium Club, patient acceptance has increased dramatically.

I would like to state that the use of Lithium Carbonate, this common mineral, has revolutionized my psychiatric practice. Lithium Carbonate as an adjunctive therapeutic tool has allowed me to formulate an approach to psychiatric distress and illness that has dramatically shortened the length of therapy for most individuals. I have been able to utilize other psychiatric approaches more rapidly and more successfully. Lithium Carbonate has allowed me to help stabilize our patients' moods. With mood stabilization, assertiveness training, behavior therapy, and classic analytically-oriented therapy, have become more effective.

Chapter Ten

CAN YOU READ BETWEEN THE LINES?

As an epilogue to this text on the multiplicity of Lithium deficiency syndromes and the overwhelming, almost universal application of externally supplied Lithium Carbonate, I felt patient reactions and discussions would be useful. I asked many patients to compile personal histories on their exposure to Lithium Carbonate. I asked that they stress the positive and negative equally.

As I expected, most of the individuals I asked, although well-intentioned, did not produce an essay or manuscript. There were excuses from, "I didn't have time," to "Whatever I have to say has already been said," to "Gee, Doc, I wrote it but it was so short, I didn't bring it in."

Several patients did produce essays, however, and many of them have been included in this section. At the end of each essay, I felt it would be useful to comment from my clinical standpoint about how I viewed my initial and subsequent interactions with these patients.

JUDY

It's a wonderful feeling to wake up knowing that the day is not going to be a crisis because my hair didn't turn out just the way I expected. Even more wonderful is knowing that my mood will be fairly predictable. Before Lithium, I never knew if a particular morning would be the beginning of a high or the beginning of a low.

Lithium Carbonate has literally changed my life. I began taking Lithium after I was involved in an automobile accident. Before the accident, I thought of myself as a well-adjusted, mature twenty-nine-year-old woman. When I felt good, I accomplished an incredible amount of work, both professionally and personally, but without a plan or direction. I was able to get by on little or no food. Conversation with friends or family could go on for hours, only I dominated the discussion and rarely "listened" to what others were saying. I angered easily, but always internalized it for fear my temper would take over, as it often did. I needed very little sleep during these highs and would toss and turn if I got my body to bed by eleven o'clock.

During a low, I constantly thought about death: how and when would mine come? how would I cope with the loss of my husband or parents? I still managed to do the work I had to as a teacher and tried to keep an organized home, but there was no enthusiasm, just resentment about having to do these chores. I tended to eat more junk-type food and felt guilty about eating it. I felt depressed and rarely felt like talking. I was too wrapped up in myself and my feelings. I still got angry, but felt guilty for having these feelings. At these low times, I felt especially unsure of my husband's love for me. How could he possibly desire a miserable, unattractive fatty?

These periods of ups and downs were so much a part of my life for such a long time that they became normal for me. I didn't think they were normal for my husband and close friends, just normal for me because I was different. I expected to feel down, then up. I knew these moods would pass; I just never knew how long they would last, or why they changed.

In June, 1980, my husband and I were involved in a serious automobile accident. We were passengers in a car that was hit by a drunk driver. I am the child of an alcoholic, so the circumstances of the accident did not help me cope with the feelings of depression that emerged afterwards. I went through my husband's hospitalization feeling numb, as if in a dream. Our lawyer suggested we both seek psychiatric counseling to help us deal with our anger.

I've been taking Lithium Carbonate since November 1980. My moods are more stable. I still have highs and lows, but the highs are much lower and the lows are much higher. I can still accomplish a great deal of work, but I am able to set priorities. If I don't accomplish a task, my world doesn't fall apart . . . I can always do it tomorrow! I still have feelings of anger, but

I am able to place them in their proper perspective and deal with them or the person who has caused my anger in a rational manner. I am finally able to listen to others during a conversation. It's amazing to discover how interesting other people's ideas and thoughts are. My speech has slowed down to half the speed it was, and the topic of death has taken its proper perspective in my inner thoughts. I'm too busy seeing how beautiful the world is! I seem to notice the most minute details about things and find beauty or humor in situations which I failed to see when I was so wrapped up in myself. I've discovered that people laugh at my sense of humor. I feel very secure about myself as a person. I see that I'm pretty and thin. It's not a crisis if my hair isn't just right. If I've eaten too much food the day before and feel like a fatty, I know I can do something about it. I don't let that feeling dominate my day. My relationship with my husband was always terrific, but since we're both taking Lithium, it's even better. I'm more honest with him about my feelings and I trust his feelings of love for me. After all, I'm a good person to love.

These are some of the wonderful results of Lithium Carbonate. There are some negative aspects, too. I've gained a little weight. I can live with it knowing that there are so many positives. My sexual relationship with my husband has changed. I've had a loss of sex drive. My sexual urges weren't nymphomaniacal before Lithium Carbonate, but since I've started to take it, I've noticed a definite loss. It's a bit of a problem because my husband's sex drive has always been much greater than mine, and I sense feelings of rejection and resentment from him. I have hand tremors, too. They have been embarrassing in many situations, particularly when my third grade students have noticed them.

Lithium hasn't made my life perfect, but it certainly has changed it. I feel like the same person, but also very different. Shortly after beginning the Lithium therapy, I told Dr. Fox that I felt as if someone had picked me up off a speeding merry-go-round and nailed me to the ground. Lithium Carbonate keeps the merry-go-round of my life on slow speed.

My first meeting with Judy was in October 1980. At that time, she visited my office upon referral by a former patient. It is important that because of my interest in detecting Lithium deficient individuals, I am acutely aware of what questions need to be asked and what observations need to be made to establish the existence of the syndrome. The first and foremost nonverbal clue is the ease with which an individual maintains eye contact during conversation. The more intense and comfortable the eye contact, the more likely an individual is Lithium deficient. Although the eye contact is almost continuous, it is not of an uncomfortable type. One does not feel stared at, but rather talked with.

Judy was heavyset. Although most Lithium deficient individuals are not, most heavyset individuals appear to be Lithium deficient. Her chief complaint was, "I'm confused. I'm running around in circles." As her scenario unfolded, she related feeling impatient and short-tempered. She described herself as moody, with feelings of low self-esteem and low self-worth. She also noted that she frequently apologized for her behavior patterns. She was a college student, but she began to discuss her early childhood experiences. I told her I would rather discuss the here and now. I recognized that she had a special interest in psychology and psychiatry, but I explained that discussion of early childhood experiences, although interesting, was not useful during an initial psychiatric interview.

We returned to discussing the here and now. She told me that she was involved in a large number of "outside activities," always willing to get involved in any activity if it "sounds good." She called herself a "volunteer type" and said she was easily influenced by the needs of her group and was willing to volunteer to help them in order to feel needed. She then recognized that she had a very high set of standards which were unrealistic and at times rigid. She also noted that she did not deal well with criticism and that she was a perfectionist. She said she worried a great deal about many things, and tried to preplan many of her reactions and interactions with people.

Most of all, she recognized that she was very verbal and had an intense need to talk.

I generally try to establish an immediate rapport with all my patients, often with the clinical ploy of challenging many of their basic tenets. In Judy's case, when she said she was afraid of criticism and was a perfectionist, I told her that she was, in fact, incorrect. She reacted with a startled response, at which point I told her that if a perfectionist walks into a room and sees twenty chairs in disarray, she gets an internal pleasure out of setting them up neatly and in order, in rows of four or five. I told her that she does not get pleasure out of setting the chairs up in a perfect pattern, but she does so to avoid criticism by others if the chairs are not neat. She thought for a moment and agreed. We therefore established a questioning rapport. She wondered how I could challenge one of her most basic tenets in life and prove in but a few moments that the way she observed her own behavior was incorrect.

I point this out so you can recognize that behavior is, in fact, primarily determined genetically. When one evaluates and understands the basic tenets of the genetic predisposition of a Lithium sensitive individual, it is not an exercise in black magic or reading a crystal ball which allows me to challenge an individual's basic tenets; rather, it is a clinical application of an understanding of personality patterns.

Because Judy was a college student and had a more than average knowledge of psychology, I decided not to present her need for Lithium Carbonate on her first visit, but to wait until her second visit.

As I have mentioned in other sections of the book, when you attempt to talk directly to individuals who suffer from a Lithium deficiency syndrome, often they either latch onto the first part of a statement and do not hear the second part, or they become so involved with the answer they are formulating that they rarely hear the full question or conversation. Judy fit into this scheme of miscommunication. The second session was very frustrating and we had a lot of difficulty attempting to relate to her and her needs. When an individual who suffers from a Lithium deficiency is on a relative high, it is like a forty-five speed record going at seventy-eight. This situation is further complicated by the fact that multiple records or tapes are being played simultaneously at this seventy-eight speed. Over the years, I have developed the technique of talking indirectly to the patient.

I have found that in order to have Lithium deficient individuals listen carefully, the best approach is not to talk to them directly. A statue of Nefertiti's bust sits in my office. Nefertiti is a Greek goddess who is well known in the history of Egyptology. I began explaining to Nefertiti why Judy needed external supplements of the mineral, Lithium Carbonate. After listening intensely for a few minutes, she became irate and overtly frustrated and began to show signs of anger. She described her displeasure: "you're ignoring me." I explained that the best way for her to listen was for me to talk about her rather than with her. She refused to accept this as valid. As a therapeutic technique, however, I was able to gain her full attention and she began to deal effectively with appropriate questions relative to her need for external supplements of Lithium Carbonate. The overall results were that after two psychotherapeutic sessions, Judy agreed to take Lithium. As you can see from her own evaluation, her agreement was superficial and cognate rather than a feeling agreement. With Lithium sensitive individuals, I can only be sure of compliance and cooperation after an agreement is obtained from their visceral computer.

Several weeks later, Judy returned to my office. She said she was not as yet taking her Lithium Carbonate. It was obvious that she was waiting for a reaction. Instead of admonishing her, I acquiesced to her control. I clearly stated, and it is a matter of fact, that each and every patient is in control of his own therapy. They either keep or do not keep their appointments; they fill or do not fill their prescriptions. If they fill them, they take them according to schedule, or they do not. I explained that there is no control issue at stake between us. It was clear that Judy was in total control. I explained that it was my function to present information to her in such a manner which would make her compliance palatable, and that whether or not she got well was up to her.

Several visits later, despite her relative lack of cooperation, she recognized that her son needed Lithium Carbonate and she brought him in for treatment. Treatment was begun. She obviously was in favor of it and would therefore comply herself.

Over the next several months, her clinical course waxed and waned. During an extensive time frame of approximately four to six weeks, she described feeling stable and was handling herself, her family, and her education appropriately. However, she decided to test her need for Lithium Carbonate and she crashed. When she didn't take her prescribed minerals, she became anxious, tense, argumentative, verbally and physically assaultive, and abusive. Only then could she recognize the need for Lithium Carbonate.

I didn't admonish and belittle her for her lack of cooperation; I reinforced it. I explained the fact that most patients at some point discontinue their Lithium Carbonate. They begin to feel as if it is unnecessary and that they are feeling good because of their natural mood and not because of the Lithium Carbonate. Consequently, most patients, especially males, discontinue their medication. She did so and she then recognized her need for it and resumed taking her mineral supplement.

Most patients are not satisfied with this lesson. They usually discontinue their Lithium Carbonate at least once or twice more within the next several months before they fully acknowledge its usefulness and its help.

Over the years, I have found that it is much more efficient to acknowledge their errors and work with them rather than to admonish and attack them for doing wrong. It has been my clinical impression from a behavioral standpoint that if you degrade patients for honesty, you are teaching them to become dishonest. That is overwhelmingly detrimental to therapy. It is much more useful to work with patients' needs and praise them for their openness and honesty rather than training them to be dishonest and devious.

We chose to present Judy as a relatively typical example of the way many patients approach the acceptance or rejection of

Lithium Carbonate as a mineral substance that is necessary. Judy was rather typical. Despite the fact that she liked to be regarded as unique, she was willing to accept the fact that personality style and actions are relatively predictable along the lines of genetically inherited personality traits. Admittedly, there is a great variation within these traits; however, the basic overview is predictable. This is important because it makes treatment much more rapid and much more efficient.

VICKY

One of the most unique experiences I can imagine is to meet yourself, your true self, for the first time and find out that you really like yourself. At thirty-four, I found out that I like myself, and I like being me. Most of my life, I wanted to be someone else, not any one person in particular, just anybody other than me. It is apparent to me now that I've spent enormous amounts of time and energy doing just that, trying to be someone else. Usually that someone else was what everyone else wanted me to be. Actually, I've tried to be what everyone I've come in contact with wanted me to be, starting with my parents.

As early as I can remember, I've had mood swings. I would be elated most of the time, so I suppose to my parents and other observers I appeared to be an unusually happy, active, well-adjusted child. Yet I would occasionally have days when I couldn't seem to do anything. I couldn't give my mother any physical reason for being "sick," so I would have to go to school and do the usual activities expected of a child. During these times, I was tired, sad, and prone to tears. I started making up "sick" excuses for my mother so I wouldn't have to get out of bed. I really did feel "sick"; I just was not able to say exactly why.

When I was eight, I fainted for the first time. The school nurse called my mother. They discussed numerous reasons why this might have happened. My mother seemed to feel it was caused by fear from the previous night. My father, an alcoholic, had been drunk and abusive. The nurse thought it was more likely that I had been injured, because I complained of great pain inside my head. I knew I had not been injured, but my head hurt so much and I didn't know why. I now believe this was probably the first of a type of "spells" I was to have as I grew older.

While I was growing up, my father was wonderful when he was not drinking. He was the most loving, attentive father anyone could want, but, when he drank, it was as if I didn't know him. He was verbally and physically abusive, mostly to my mother, sometimes to me and my brothers and sister. I loved him so much when he was not drinking, and was terrified of him when he was. I spent most of my grade school years moving back and forth between living with my father and living with my grandparents. We would go there when my mother would have to leave him. I should mention that my father did not get drunk consistently. He would go for weeks, sometimes months, completely sober.

When I was eleven years old, I "fainted" for the second time. This time I was taken to a doctor to be examined. I had the same type of pains in my head. My parents were terribly concerned, not only because I fainted but because I fainted in the traffic lane of a busy street. After a brief examination, the doctor assured them that the only thing wrong with me was "growing pains." The episode was closed, to my parents' relief, by the "wise" doctor.

When I "fainted" the second time, I don't think I was unconscious in the usual sense. Granted, I had no conscious awareness of my surroundings, but I did have sensations. I experienced a beautiful sense of warmth, peace, and tranquility. These were feelings I had not experienced before. My life thus far had been more akin to a razor's edge, sharp, intense, and so very hard to cope with. But what could I do about it? I was only eleven years old.

When I was twelve, our house caught fire and was completely destroyed. No one was injured, but I think it was this incident that consolidated all of my fragmented fears into a fear of everything. It was also the beginning of puberty, which is difficult for almost everyone, and it was the beginning of junior high school, another unexpected trauma.

My family lost everything we owned in the fire. The only clothes we had were from other families in town who gave them to us. Just before the fire, my parents had separated for the last time, and my mother was supporting herself and four children, me being the oldest. Being part of a poor family in a wealthy town is difficult in itself. Being twelve, entering a new school where you don't really know anyone, definitely adds to a feeling of insecurity. When girls started coming up to me, sneering and pointing out that I was wearing clothing that had belonged to them, I was totally embarrassed. Unfortunately, that was only the beginning.

Have you ever noticed that every school has its "scapegoat," that one person who just about everyone picks on? Well, I was elected to be that person in my junior high school. For two years I was criticized for everything I wore, said, and did. I was tripped in the halls, my books would go flying off my desk, my gym clothes would be missing, etc. I was everyone's end for whatever practical joke was making the rounds.

The more I was subjected to this type of treatment, the more shy and withdrawn I became. The shyer I became, the more I received this type of treatment.

When I entered high school, I joined another group of outcasts and nonconformists called "hoods," short for hoodlums. These were the school's troublemakers. Even with them, I was only on the outskirts of the group, but at least I felt I belonged somewhere. I had other kids to talk to, someone to eat lunch and walk to classes with.

I had always been an excellent student until I reached junior and senior high school when my grades dropped to average. The attitude of the "gang" I was involved with was "Who needs school?" Most of them were barely passing, so I didn't want to do too well. After all, only the "popular" kids got good grades and ran for student council. I was neither fish nor fowl. I needed to belong somewhere, anywhere.

When I was old enough to date, I always dated people who weren't from my school. Usually I went out with "older" men in their twenties. I would have liked to date some of the boys in my school, but no one ever asked me. I was still very shy around my classmates. I could be friendly and feel more adequate only by meeting someone from the outside.

When I was almost eighteen, I became engaged to a man of twenty-four. My parents didn't like him or the situation, but there was very little they could do about it. I was very stubborn about seeing him. "Here," I thought, "I've finally found someone to love me, and for me to love." I felt I had so much love to give and until then, no one to give it to. Chris was smart, loving, considerate, and mine. I was going to be the best wife, and eventually the best mother ever. No one could tell me Chris was no good; I knew better. But after a three month engagement, Chris disappeared. I looked for him for a month before his sister finally told me he was in jail but he didn't want me to know.

The feeling of desperation turned into depression, and I started dating David, the man who would eventually be my first husband. For some reason, I believed the reason I lost Chris was because I had not given in to all of his demands. I resolved that this time it would work out because I would give whatever was necessary. The evening of my senior prom, I slept with David. Afterwards, I didn't feel very good about having done it, but I rationalized that since most of my friends were no longer virgins, at least I had become one of the crowd.

Graduation was only a month away, and my best friend, Carol, and I had been planning on getting an apartment. We moved to New Bedford, Massachusetts. David just happened to be stationed at Otis Air Force Base, only a short drive away. This was really going to be an adventure. I'd be on my own for the first time, near the man I "loved." Carol and I found an apartment, but we didn't find jobs. Yet, we were sure that was only a temporary situation.

Nothing worked out as I had planned. David and I "broke up" about a month after Carol and I moved to Massachusetts. Several days later, Carol took me to a doctor. I thought I was terribly ill. All I could do was throw up and sleep. I was not sick, I was pregnant, and terribly, terribly frightened.

In one sense I was lucky. David was a very concerned, sensible individual. When he learned I was pregnant, he said he would stick by me and help in any way he could, short of marriage, that is.

Eighteen, pregnant, depressed, scared, unwed, and unloved. Even now, I hurt inside when I remember. I couldn't possibly tell my family back in New Jersey. They were very straight and church oriented and thought this situation just didn't happen to "good girls." I felt so alone and guilty.

The apartment Carol and I lived in was on the third floor. One night, I crawled out of a window onto the roof and tried to make myself jump. I didn't really want to die; I just couldn't live in my situation any longer. If I jumped, I might not die; I might just end up mangled for as long as I lived. I decided to try to cope.

David took me to an adoption agency to discuss the procedure with a counselor. He thought this seemed to be the best option. I was doing very little thinking for myself at this time. All I could do was feel, and all my feelings led to depression and fear. Part of my mind was still denying that I was pregnant: All this was a bad dream, and eventually I would wake up.

I managed to hold myself together until December, which was my seventh month. I had been making excuses to my family about why I hadn't come home to visit since August. I had gotten a job as a waitress in a coffee shop and I told them I couldn't get off for the holidays. I think they believed me, but I'm not sure.

Carol, David, and Bob (David's best friend) told me they were all going home for Christmas. I don't know what I expected of them, but I hadn't thought of being all by myself for the holidays. After they left, I was even more afraid and depressed. What if something happened and I had the baby when no one was around? I had no phone and no car. What if the baby came and we both bled to death? What if I fell down the stairs? What if the building caught on fire? What if, what if, what if? I was so alone and nobody cared. I had to go home. I needed to talk to my mother. Even if she never forgave me, at least I wouldn't be alone anymore.

I packed my suitcase, walked to the bus station, and bought a ticket to Philadelphia. I was doing something; at least I was doing something. The next bus wasn't until nighttime, so I waited. That wasn't nearly so hard now that I was doing something. I was going home.

The only bus I could get left me with a three hour stopover in New York City. I fell asleep in the waiting area while I was waiting for the next bus. When I awoke, I discovered someone had stolen my wallet, so I had no money except for the change in my pocket. At least I still had my bus ticket, so I made it to Philadelphia, which was that much closer to home.

I didn't have enough change for the short bus trip from Philadelphia to my hometown in New Jersey. It didn't really matter at that point because I had gotten a case of "cold feet." Instead of going home, I took a bus to my friend Barbara's house, within my bus fare range. She was quite surprised to see me, and was quite curious about my condition. After I had explained what had happened over the past months, she volunteered to call my mother since I was too afraid to call myself. Barbara told my mother she had a Christmas surprise for her from me and that she would pick up my mother and take her to it. My mother agreed and they arranged to get together that evening at seven o'clock.

That was such a long day, and yet not long enough. I wanted to see my mother but I was afraid. In my mind I went over and over what would happen when she got there. Seven o'clock finally arrived and my mother walked in. She didn't care about my condition, only that I was safe and with her.

I cried and cried and told her about all the hurt, frustration, and emotional pain I had been suffering. She told me she loved me no matter what, and that we would work everything out. The only thing she said that frightened me was that my father would have to be told.

Even though my parents were divorced, my father lived only a short distance from my mother. She called him, briefed him on the situation, and arranged for him to meet us at her house. By this time, I almost wished I had stayed in Massachusetts and braved the loneliness.

By the time we reached my mother's home, my father had worked up a pretty good head of steam. First, he wanted to see the man who had gotten his little girl in trouble. He asked me if we were getting married. When I told him "No," he told me that if I loved him enough to sleep with him, I loved him enough to marry him.

David didn't want to get married. By this time, I realized I didn't want to, either. I only wanted a way out of the situation I was in without getting hurt any further and without hurting anyone else.

By the time everyone who felt entitled to expressed their opinions and the votes were counted, it was decided that David and I would have to get married. I was so mixed up that I agreed and David had little choice, considering the pressure he was under, so he agreed also. After the holidays, we drove back to Massachusetts, reluctantly obligated to get married.

The baby was due on February 17. We didn't have a lot of time, but we procrastinated. Finally, Bob took matters into his own hands and started making arrangements. I think he had visions of a labor room wedding and was trying to avoid that. He took us to get the license, called a church, and set a time.

On January 28 we were married. It was so hard to say the marriage vows. I felt as though I was lying to God and everyone else when I made the standard promises that I knew I wouldn't be able to keep. I thought of running out of the church before it was too late, but I knew I couldn't. I didn't want to make more of a fool of myself and there was nowhere to run, anyway. I felt so lonely, so abandoned, so very alone.

After the ceremony, we only had enough time to get something to eat before David had to take off on a three day flight mission for the Air Force. I was very pregnant, finally married, but still alone. I may not have loved David, but he was all I had and I wanted him to be with me. I had resolved to try and make the marriage work. I thought even a bad situation might turn out okay if I worked at it hard enough. After David came back from his flight and we settled into base housing, the situation didn't improve. One day he would try very hard to be nice to me, and the next he would be silent and angry. We had no communication at all. I never knew what he was thinking or what he felt. When I would ask him, he wouldn't answer.

On February 14, I started having labor pains about eleven o'clock at night. David said to let him know when I wanted to go to the hospital and he went to sleep. After a sleepless night of pacing around the house, picking up things and putting them down, trying unsuccessfully to read but mostly watching the clock, I woke him at six o'clock when the pains were four minutes apart. He drive me to the hospital, checked me in, and left. I didn't even know he left until a nurse told me. I still don't know if I have ever felt so little, scared, or helpless as I did then.

A woman brought in some papers for me to sign, one of which was permission for the anesthesia, a spinal. When I asked her what a spinal was and she explained that they stick a needle in your spine at the waist level, I told her I didn't like the idea. She told me I could take it or leave it, that it was all they used, and my only other alternative was nothing. Well, I certainly didn't want nothing, so I signed. She lied to me, but I didn't find that out until later. I suppose getting me to sign the form was easier than having to make out a new one. I didn't realize that I was the type of person who didn't want to displease anyone!

There was one nurse, to whom I'm very grateful. When she found out that I was alone, she made every effort to stop in and stay with me as often as she could. She was the only person who kept me from sheer panic. I wish I knew where she is now so I could tell her how much she meant to me on that day.

My son was born at 2:18 p.m. on February 15. The hospital was going to notify David, but I didn't know where he was and he hadn't called in since he dropped me off that morning. I wasn't allowed to see my son right away because they told me they had to drain the fluid in his lungs. I finally got to see him late the second day. He was eight-and-a-half pounds of wiggles. The nurses told me he had the best set of lungs for crying in the nursery. I was afraid he cried so much because the fluid had hurt him. He would sleep for fifteen minutes and then cry for an hour. He didn't seem to have any schedule other than that.

Because my temperature was elevated, they wouldn't let me come home for eight days. David drove to New Jersey and came back with my mother and his mother. My mother stayed for five days before she had to go gome, but she sent my sixteen-year-old brother, Bret, up by bus. She told him, I found out later, that she didn't like the way I was acting and was worried.

Bret was with me for about a week when I came home from shopping, parked the car in the carport, and couldn't get out of the car. Each time I tried to stand I would have shooting pains run down my legs. I partly walked and mostly crawled toward the front door. I got about half way when I started to lose the feeling in my legs. I was terrified and I started screaming. Bret came running out and carried me into the house. He called an ambulance and made several phone calls to try and find David, since he and his friend, Bob, had gone out for the evening.

The ambulance took me to the base hospital and Bret stayed home with my son. By the time we reached the hospital, I was hysterical. I didn't know what I was saying or doing; I only knew I was in pain. The doctors kept trying to talk to me, but I couldn't understand what they were saying. It was as though they were speaking in a foreign language. I can remember faces close in front of mine, faces I didn't know which kept shouting words at me that I couldn't understand.

Someone must have given me a shot because I fell asleep. When I awoke, my "husband" and one doctor were present. I could understand them when they asked me to tell them what was wrong. I explained the pains and numbness. The pains weren't nearly as bad as they were earlier, but the numbness was still there. The doctor had me admitted and ordered X-rays of my spine and an electrocardiogram.

I was confined to my bed and wrapped tightly in ace bandages from the waist down. Three times a day a nurse would unwrap me, rub my legs for a few minutes, and wrap me up again. There were several different doctors who stopped in to talk with me. After they would leave, I'd usually have a new test to undergo. I've always wondered, "Did they care, or did they do what was clinically expected of them?"

After around five days, I was allowed to go home. My legs seemed weak but otherwise fairly normal. I was told I had had an acute bladder infection and that had caused my problem. I didn't believe the diagnosis, but they had to tell me something. A doctor can't very well tell you that he doesn't know what's wrong.

About a week after I came home from the hospital, Bret had to go back to New Jersey. David and I and our son were now alone for the first time.

I couldn't understand David most of the time. He was always so moody and silent. I know he hated being married so I tried everything I could think of to make it easier and more pleasant. I cooked his favorite meals, kept the house in good order, and tried to keep the baby out of his way. Our son cried more than he slept, and that seemed to infuriate my husband.

The more I tried, the less response I received, and the more frustrated I became. The house became a mess, I cried almost as much as the baby did, and David went back to acting like he was single. He would go out most nights with his buddies and on occasion not come back for a day or two.

When our son was about six weeks old, David and I had a really furious argument. He was leaving on a three day flight mission and we fought right up to the moment he left. I begged him to try to understand how I felt, but I guess he either couldn't or wouldn't.

After he left, I was crying and completely frustrated. I tried to call my mother in New Jersey, but there was no answer. The baby was crying again, so I checked to see if he needed changing or if he was hungry. He appeared to be fine, but he wouldn't stop crying. I tried walking him, but he cried even more. I walked him around and around the living room, both of us crying and not being able to stop.

I couldn't bear all of his crying. I put him back in the bassinet and turned on the television, the stereo, and the radio as loud as they would go. Even with everything full blast, I could still hear him crying. It was like screams echoing through my mind.

I took a pillow and forced it against his face. I know I didn't want to kill him; I just had to stop that screaming, no matter what.

I don't know how long I stood there. Suddenly, I realized that he would suffocate from the pillow. I tore it away and threw it across the room. I checked him over carefully and found him to be all right, just exhausted from all of his screaming.

Then I realized what I had done. I had almost killed my son. I didn't mean to, but the result would have been the same.

I must have snapped at that point because I don't remember anything until someone was slapping me in the face. Bob told me he had come in when he heard all of the noise and no one answered his knock on the door. He said he found me beating my head and fists against the wall. My forehead and fists were bruised and bloody, but I don't remember doing it at all.

Bob cleaned up my face and hands, and kept asking me what had happened, but I couldn't seem to talk very well. I didn't even feel like trying. I just didn't care anymore. Everything was senseless. I just wanted to go to sleep and not have to wake up. I was so tired and I badly needed to rest.

I told Bob to go away and leave me alone, but he wouldn't. He called the base chaplain and arranged to bring me right over. He left my son with a neighbor and drove me to the chaplain's office. After Bob talked to him for a few minutes, the chaplain called the base hospital and spoke to the base psychiatrist. I was then taken to the hospital to see him. He was one of the doctors who came in to "chat" when I was there recently. I must have said something to him, but I don't remember what it was. I do remember that he told me I was very sick and needed several years of psychiatric treatment. This frightened me terribly. I didn't really feel sick; I just couldn't find anyone to understand how hurt and lost I felt. He also told me that my mother didn't love me as much as I thought she did. That was the last straw. My mother was the only person I knew I could trust as far as loving me was concerned and no one was going to say that. The doctor made a lot of appointments for me to come back and to see him. He told me to go home and try to relax.

I did go home but I didn't relax. I started packing my things. I was going home to live with my mother and no one was going to stop me. Instead of just trying to cope with the situation, I was going to do something about it.

When David came back from his flight, I told him my decision and asked him to drive me back to New Jersey. He said it was going to have to be that way anyway because he had applied for overseas duty and had been accepted for an eighteen month remote tour.

I picked up the baby's and my hospital records to hand carry to Walson Army Hospital, the nearest military hospital to my hometown, and I found out what the base psychiatrist had written about me. He said I was psychotic and likely a paranoid schizophrenic with masochistic tendencies, and strongly recommended extensive psychotherapy. I remembered my father was diagnosed as a paranoid schizophrenic when I was a teenager. He stayed in a mental hospital for awhile and received shock treatments that were supposed to cure him. No one, but no one, was going to do that to me. I vowed I was going to get better on my own and never tell anyone about this situation.

I moved in with my mother and David left for overseas. I was relatively comfortable living with my mother and raising my son. I had decided that the best thing for me was to get a divorce and try to make a fresh start. I found a job in an office and settled down to learning how to earn a living. This was my first real job, not counting being a waitress in the coffee shop. At the office, I was learning new things, meeting new people, and beginning to feel like a real person.

David came home on leave when the baby was almost a year old. He spent most of his time with his family and friends (he was originally from the same general area). He did spend some time with me, however, just enough it seemed for me to get pregnant. Surprisingly, I found myself rather happy about having another child. I was really enjoying my son as he was growing, learning to walk, smile, and play. I was hoping I would have a little girl this time.

When I was three months pregnant, I started having pains one day while I was at my office. I went to the ladies' room and realized I was bleeding. I had a miscarriage on the floor of the ladies' room. I was taken to the hospital and had a D and C (dilatation and curettage).

I returned to work a few days later, thinking it would be better to have something to do rather than sit home feeling sad. About two weeks after the miscarriage, a man who came into the office every so often asked me what was the matter. I told him I was sad because of personal reasons. He said he had the perfect cure: why didn't he pick me up after work and buy me a drink. He seemed like a nice sort of guy so I agreed.

He took me to a little bar and we had several drinks. I never drank much, so several drinks made me feel pretty loose and outrageous. He suggested we go to his apartment to "relax," and it seemed like a good idea at the time. I suppose the feeling of having someone pay attention to me for a change made me stupid. I didn't realize what his definition of "relax" was until we got to his place. He wanted to play around. I said no and he called me a tease and said I deserved what I was going to get. I was really frightened. He wouldn't let me leave. Physically he was much stronger than I was, so he had his way. The worst damage I suffered at the time was a black eye and a bruise on my arm.

When I got away, I called a policeman and told him what happened. He said it sounded as if I had led the man on and he doubted if a judge would do anything about it. Besides, did I want my name plastered all over the newspapers? Instead, I called a friend and had her drive me to the hospital to see the doctor who had done the D and C two weeks before. He cleaned me out and said the odds of getting pregnant were a million to one (this being my main concern) since I had just had the D and C. He said if I didn't get my period in six weeks, I should come in to see him and he would give me a shot to bring it on.

I didn't get my period, the shot didn't work, and I tested pregnant. The day I got the results of the pregnancy test, I also received a letter from my husband. He said he was sorry he had been so hard on me, that he wanted to try to make it up to me, and that he thought that if we worked at it, maybe we could have a good marriage. He also told me that he loved me, for the first time since I had known him.

I quit my job because I couldn't stand the idea of running into that man again. I waited out the month until David was due home on leave. While I waited, I explained to my doctor exactly what had happened. Most military doctors have a reputation for being cold and unfeeling, but this man was very warm and sympathetic. He arranged for me to go before a board of doctors to determine whether or not I would be allowed to have an abortion. They decided I could, if my husband agreed.

When David came home, he was so loving and kind I couldn't tell him right away. I procrastinated for a week. I knew I had to tell him, but I didn't know how, and I didn't want to ruin this new relationship we were developing.

When I finally told him, he was very understanding. He said having an abortion was my decision. I told him I didn't think I could go through with one since I had already felt life. He said he didn't want to raise a child fathered by another man, so we decided adoption was the only way out of the situation.

David returned to his overseas tour of duty, and I waited out my pregnancy. After the birth of my first child and the depression that followed, I seemed to have chronic depressions with brief periods of elation in between. During this pregnancy, I was in a constant state of depression which just kept getting lower and lower. I craved salt. Everything I ate was well salted. I would even go to the kitchen and pour about a teaspoonful of salt into my palm and eat it. The more salt I ate, the more salt I wanted. I gained seventy-three pounds during this pregnancy. I told the doctor I thought the weight was fluid, but he said it wasn't because I wasn't carrying any fluid in my limbs. Instead, he gave me diet pills. I didn't loose any weight, but I felt terrific for the first time in years. The more pills I took, the better I felt, until I ran out of pills. Then the depression was even more devastating. The doctor finally put me in the hospital on an eight hundred calorie a day diet. After eight days, I had lost four ounces. He finally gave me a diuretic, but the next day I went into labor and delivered another son.

The day after I delivered, a social worker came in to get me to sign the papers for the adoption agency. I said I wanted to see my son first. He said that was impossible. The mother of a child being adopted was never allowed to see the child. I said that was ridiculous and that I would sign nothing until I saw my son. After much argument with doctors, nurses, and social workers, they finally agreed to let me see him.

They took me to a private little room and brought him to me. I named him Michael, for the time being, at least. I talked to him and explained why I was giving him up. He was the most beautiful little baby. I wished I could have kept him, but I knew my husband would never have accepted him and he would be better off with two parents who loved him.

David came home from overseas and was restationed in Florida. Two days after I came home from the hospital, we left for Florida. The first thing I did when we settled in was to go to a doctor because I was depressed. He said it was just postpartum depression and that it would go away in time. I tried to talk him into some of the diet pills the other doctor had given me, but this doctor said I wasn't overweight and didn't need them. I couldn't very well tell him the only reason I wanted them was so I could feel better.

At first, David made an effort to get along with me. He couldn't understand why I was so depressed all the time. I don't blame him because I didn't understand it either. I tried and tried to be better, but nothing I did worked. Eventually, we evolved back into the same type of relationship we had in Massachusetts, which only added to my depression. I started having occasional periods of elation, but these were few and far between. They were also frightening. I found myself talking incessantly and not being able to stop. We argued most often about the phone bill: I was constantly calling my family and friends in New Jersey, not for a specific reason, just to have someone to talk to. I had an almost constant feeling of desperation.

When David's tour of duty was up in Florida, he got out of the service, which was a great relief to me. We could go back to New Jersey where there were people who loved and cared about me.

David got a job with RCA and we purchased a house in Easthampton. I was sure everything would settle down and that we would be happy. David still remained very silent, but stayed home and didn't run around with his friends.

I set out to be the happy homemaker type. I enjoyed cooking and cleaning, and the wife and mother roles. I became the "supermom" of the neighborhood. By ten o'clock in the morning, my housework was done for the day and I was free to do the things I wanted, except I didn't know what I wanted to do. The more projects I started and didn't finish because I lost interest in them, the more frustrated I became. The more frustrated I became, the more depressed I felt. I finally went out and got a job. Along with the job, I found a source of diet pills in a co-worker. Now I would feel happy again.

Well, I did feel happy for a while, but when the pills would wear off, things seemed worse than ever.

My father, who was still an active alcoholic, would come to my home when he was drinking and would be abusive. I was still conditioned from my childhood and would react with the same type of fear I felt as a child. Several times, I had to call the police to have him removed.

One day when I had taken a diet pill and was feeling pretty good, my father came over and was drunk. During the battle that ensued, I locked myself in the bathroom and became hysterical. I just screamed and screamed and refused to come out. My father sobered up somewhat and became quite alarmed. He called a doctor and asked him what to do. The doctor asked to speak to me. I said I would speak to him if my father would leave the house. My father said he would, so I came out and spoke to the doctor. He asked me if I had transportation, and when I told him I did, he told me to come straight to his office. Once there, I explained to the doctor (a Doctor of Osteopathy) what had been happening, and he gave me a pill to relax me. After I was there for about fifteen minutes, I heard my father's voice in the other room. He was drunkenly insisting on seeing me. He was told that was impossible, that I was resting. He threatened to break down all the doors until he found me, that no one was going to keep him from seeing his daughter. The doctor gave my father two options: he could leave right away under his own power, or the doctor would call the police and they would remove him. My father took the first option and left.

When the doctor came back into the room where he had left me, I was curled up in a ball in a corner, unable to speak or move. He gave me a shot to calm me and with the help of another man I didn't know lifted me onto the table. A nurse was told to stay with me and the doctor left. When he returned, I was much calmer and could respond to him when he spoke to me. He said he wanted to call an ambulance and have me taken to a hospital for observation. All I could think of was the Massachusetts doctor's diagnosis. If I went to a hospital, they might want to keep me and give me shock treatments. I insisted on going home. The doctor made me rest there for about an hour before he could let me drive. When I left the office, my father was waiting down the street. I was afraid to go home. My son was at my mother's house, so I drove there and stayed.

This doctor became my family doctor. I found he would write a prescription for me for almost anything if I gave him a plausible excuse, such as, I was quitting smoking and needed diet pills so I wouldn't gain weight.

After I had gained some trust in him, I tried to tell him about the "spells" I had been having since my son was born. They usually followed a high, elated period and led directly into an extreme low and depression. I say "usually" because there was no regular pattern. At times, they would come on without warning. I would have what felt like an electric shock shoot up my spine and explode in the back of my head. I would become hot and flushed all over at first, and then I would be freezing cold and shivering. My heart would pound and my pulse would race. My legs would be weak and refuse to support my weight. My stomach would be nauseated and my bowels would turn to water. All of this took about thirty seconds. Then what I call the "psychological crazies" would start. This is the hardest part to explain. It wasn't really like dying or being afraid of dying, it was more like being trapped in an infinite situation of ceasing to exist. Each time I had one of these spells, it was a continuation of the last one. At first, I really did think I was dying, but over the years, when I didn't die, I concluded that I was just crazy. This theory didn't help much when I was having a spell, but it did help me rationalize them the rest of the time.

My family doctor didn't understand what I was talking about. He thought the spells stemmed from a chronic state of depression, so he put me on an antidepressant. I tried to take the pills for almost a week but all they did was make me sleep. I told this to the doctor and he said to keep taking them, that I'd get used to them. I didn't, so I stopped. By now the different pills I had been taking were affecting my performance at work. One day I would go to work bright eyed and active on diet pills, and the next I would be falling asleep at my desk. I am surprised I wasn't fired.

Things were no better at home. For New Year's Eve that year, David was going to take me to New York City. I had gotten a new outfit and was ready to go when he said he didn't feel like going out. I was furious and announced that I was going out by myself. A girlfriend who lived in Philadelphia was having a party, so I went there. There was a great deal of drinking and smoking grass. I had never smoked before, so I thought it was a good time to try. Everyone said it would get me out of my depression and make me feel great. Well, it certainly didn't make me feel great. I experienced the same type of psychological crazies I did when I was having a "spell," only much, much worse. When I finally got home two days later, I was so frightened and confused I decided I really needed help. I talked extensively to the doctor at work and he said I should see a psychiatrist. He gave me a medical leave of absence and asked me to call and let him know what was happening.

I went to see my family doctor to ask him to refer me to a good psychiatrist. He said he thought all I needed was a good divorce lawyer, but if I really wanted one, he would refer me to a psychiatrist. He said he thought I should go into the hospital for a rest. I agreed and was admitted the next day. During the week I was there, they ran blood sugar, urinalysis, and blood tests. Other than that, all I did was rest. When my doctor came to see me, he gave me the name and phone number of a psychiatrist. I started to see him as soon as I came home from the hospital. In the beginning, I was seeing him a couple of times a week. All we did was talk. He had me tell him all about my life, from childhood to the present. He also gave me several types of antidepressants and mood elevating drugs to try. I was very worried about my spells, but he seemed totally unconcerned. He wouldn't believe me when I told him I only smoked grass once, on New Year's Eve. He believed I had taken acid and wouldn't admit it to him. He insisted I had experienced an acid trip, and that my previous spells were flashbacks from previous acid trips that I wasn't willing to admit. I could not persuade him otherwise.

I went to see him for ten months. When the insurance benefits ran out, he asked me how I was going to pay him. I said I didn't know that the insurance had run out and I really didn't have any way to get that kind of money. He told me if I really wanted to get well, I would find a way. The last time I spoke to him, he said if I worked out a way to pay for my therapy, I should call him.

Instead, I took matters into my own hands. I moved out of the house in Easthampton with my son, our clothes, and my car. I left everything else to my husband. I moved back in with my mother and found another job. I started dating and tried to lead a "normal" life.

When I was married, David would tell me I was oversexed. When we were getting along fairly well, the most we had sex was once or maybe twice a month. When I wanted it more, he told me I was oversexed and unreasonable. When I started dating, I decided that if my husband wouldn't sleep with me, someone else would. I needed the loving, touching, and caring. Unfortunately, somewhere along the line I mixed up loving and caring with purely physical sex. It was all so empty, but I kept on trying. Possibly I thought if I met enough men, one of them would love and care for me in the same way I wished to care for someone. I did find out one important thing: I wasn't oversexed; I was starved for affection. My career as a loose woman lasted for almost two years; I found sex without love to be empty and more harmful than good.

I found that when I was in a high elated mood I would pick a man who fit in with that mood. Unfortunately, when I would then slide into a low depressed mood, he wouldn't be able to cope with the "different person" I had become. It came down to this: whenever my moods changed, so would the man I was seeing. Now that I am better, I can see the even changing, "constant" pattern which I was weaving all those years.

When I stopped being a loose woman, I decided I would be more likely to find happiness and security in "things" than in a partnership with a man. I set out to acquire by my own merits all the material things I wanted. I was sure this was the way to be happy.

I worked at several jobs over the next several years, each one better and more responsible than the one before it. I still had mood swings in my personal life, but they didn't seem to interfere with my capabilities at work. Eventually, I was making more money than I could spend, had moved into a really nice two-story apartment with a fireplace, bought a gold Thunderbird, and everything else I thought I wanted.

I had to go into the hospital to have a minor operation, and when I came home, I hit the lowest depression ever. Here I was, with everything I thought I wanted, and I still wasn't happy. I wasn't even content. I was lonely. I had no one to share with, no one to be close to, no one to make me feel real.

I quit my job immediately. I sold my car, moved out of my apartment, gave away most of my clothes and jewelry, moved into an old house that my mother owned, and lived off my savings account to pay the bills. I mentally locked the doors on the house and told everyone to stay away from me. My son spent a good deal of time at my mother's house because I just couldn't seen to find the energy to do what was necessary to take proper care of him.

A friend of my brother's starting coming over to visit. He seemed like a warm, sympathetic person so eventually I started dating him. I found out what made him seem so warm and sympathetic: he was taking drugs. He wasn't much different than I. He described having mood changes much like mine. He said taking drugs helped him. I was willing to try anything to get out of my depression, so I started taking amphetamines. The powder type worked even better than the diet pills I had taken years before. The whole group of people I started to associate with was into one kind of drug or another. Most of them were shooting "speed" directly into their veins. I was afraid of doing that, but after a while it didn't seem so bad, and the effects seemed so good. I finally had a friend stick a needle in my arm. It was great until it wore off, and then it was unbelievably bad. I was much worse than before I tried it. Instead of staying away from it then, I went back for more so I could feel good again, and again, and again, and again. Months went by and I couldn't get off this merry-go-round. I wasn't even sure I wanted to. I didn't want to go back to being depressed all the time, but I couldn't handle the speed, which was much like my mood swings, only more intense.

I finally crashed so hard one night that another shot didn't bring me back up again. Someone gave me a cigarette with marijuana in it and said it would bring me back to normal. I was desperate, so I smoked it. It didn't bring me back to normal. Quite the contrary, I "tripped" worse than the previous time I had smoked. It was one of the most frightening experiences I've ever had. It was probably a blessing in disguise, because it scared me enough to get away from the drugs and the people with whom I was associating.

This experience left me a very shaken, depressed, lost individual. I was even afraid to sleep at night without a light on in the room.

Shortly after that, I met a very nice man. He was kind, attentive, and concerned. He listened attentively while I told him everything. He was just what I needed at that point, someone who cared whether I lived or died. He took care of things for me that I couldn't handle myself. He was in charge and I was submissive.

As I began to come out of my deep depression, I realized I'm not basically a submissive individual. This began to cause strife in our relationship. There were things I wanted to do that did not necessarily include him and he disliked this intensely. He had become very possessive of my time and resented anything that interfered with what he felt was "his time." I can't exactly blame him for becoming confused. The person I presented myself to be when we met was not the person he was involved with at this point in time. I still liked him very much, but I couldn't be what he wanted me to be.

I was still unsure of what I wanted to be or the kind of person I really was when I started seeing Dr. Fox.

I approached Dr. Fox in much the same manner as I had approached each of the other psychiatrists I had seen. I had been referred to him by a psychiatrist I had been seeing for about nine months but who could no longer see me.

Dr. Fox didn't buy my approach, nor did I buy his. I found him harsh and pushy. What gave him the right to tell me what my life was like and what was wrong with me after a five minute conversation. Who was he to even attempt to analyze my thirty-three years of living in one office visit. The other psychiatrists I had seen spent months going into all facets of my past and present. By the end of the first session, I was more than ready to leave. I had absolutely no intention of ever coming back to see this overly confident, egocentric doctor.

Just before I left, he said something that really made me stop and think. He said, "Why go on suffering? I can make you well and happier in just six months." He also said he knew that at that moment I had no intention of ever coming back, but that I would eventually, and most likely I would be desperately ill when I did. I left the office feeling completely off balance.

I went home and thought for several days about everything he had said. Many of the things he told me about myself most of my friends wouldn't know. I began to wonder if he could read my mind. It was an altogether frightening feeling to have a perfect stranger know so much about how you think and feel. I was afraid to go back, but my mind kept going back to what he said about making me well in six months. Six months wasn't really that long to invest, if it really worked. I had tried just about everything else anyone had suggested, from vitamins to yoga from friends and numerous sorts of medications from numerous doctors. Maybe I should go for one last try at becoming "normal," so I called and made the appointment.

When the time came, I was really frightened. I wasn't sure I could face that man again. He was so intimidating and I was so easily intimidated. With a little pressure, anyone could get me to do almost anything. When he showed me into his office, I was really surprised. He didn't seem to be the same doctor I had met just two weeks before. Maybe he realized how nervous I was about coming back. Whatever his reasons, this time he was gentle and sympathetic. I don't remember what we talked about, but after a while I began to feel more comfortable. The more we talked, the more I began to feel this doctor just might be the one I could come to trust.

I don't remember how long I had been seeing him when I realized that this Lithium deficiency he had been talking about meant I would have to take Lithium every day to get better. Because of my previous experiences with drugs, I didn't want to have anything to do with them much less have to take them every single day. Dr. Fox very patiently explained that Lithium was not a drug, but a mineral. I still wasn't receptive.

After I had been seeing Dr. Fox for about three months, I decided I trusted him and his judgment. I realized he really cared about people and whether or not they got well. When a previous psychiatrist stopped seeing me as soon as I ran out of money, I developed the opinion that "they" were all in it for the money. If the patient also got well, it was a side benefit and the patient got lucky. I had a very hard time believing that I had happened on a truly dedicated doctor.

It was my faith in Dr. Fox that convinced me to try the Lithium, although I really didn't believe it was going to work. He gave me a prescription toward the end of October 1980.

When I went back on the sixth of November for my next appointment, I still hadn't had the prescription filled. I'm not sure why, but I was certainly sorry. When I admitted I hadn't been taking the Lithium, I was told that I would take it and was not to miss a pill or the doctor and I would part company because he could not help me if I didn't take the Lithium. That really worried me. Dr. Fox had helped me so much already by helping me understand myself better that if I had to take those pills to keep him as my doctor, I would take them. When I left that day, I went straight to a pharmacy and purchased my first step toward truly getting well.

The first week I was taking Lithium, there were no noticeable effects. I was taking two a day. I was relieved to find out they didn't seem to hurt me in any way. I was concerned because I had heard that Lithium is radioactive and poisonous. I never asked Dr. Fox about this. I think I was afraid of insulting him by questioning his medication.

The second week I was taking the Lithium, I started having mild side effects. I was very tired all the time and very thirsty. By the third and fourth weeks, the side effects had gotten more severe. I could be standing perfectly still on the solid floor and it would feel as if it was rolling like the swells on the ocean. The walls would seem to go in and out as if they were breathing. I was very, very tired all the time and I was constantly nauseated. I had gained about ten pounds, and I had hot and cold prickles on my skin, especially around my face, neck, and shoulders. My hands trembled, and it was difficult to do any fine work like writing or drawing. The side effects were quite bothersome and a bit frightening, but bearable. Dr. Fox was very supportive, and I felt reassured.

On December 13, the side effects reached their worst. All the effects I mentioned were intensified and I was having groups of the "spells" I had been having off and on for most of my life, except now I was having several "spells" each day. It seemed as though I would just start to come out of one when I would find myself sliding into another. I reached the peak of my endurance and could take no more. I blamed the "spells" on the Lithium. I was disoriented and terribly frightened.

Instead of calling Dr. Fox to tell him what was happening, I just stopped taking the Lithium. I did call his office to cancel my future appointments. I stayed in bed for several days and retreated from the world. I had flu-like symptoms and just wanted to be left alone.

A couple of days before Christmas, I started taking the Lithium again. I don't know why. I had no logical reason for doing it; I just did. I wasn't really taking it regularly, just when I happened to think of it.

About New Year's Day, I realized I was feeling a little bit better, not much, but there was a noticeable difference. I thought maybe the Lithium was working after all. I decided to give it a fair chance and started taking it faithfully at the same time every day.

Each week I felt better than I did the week before. I started making appointments with Dr. Fox again. He pointed out to me that I seemed much improved. Until then, I thought I was the only one who would have noticed a difference. I discreetly tried asking family members and friends if they had noticed any change. The majority of them answered in the affirmative.

I was feeling better every day, both physically and mentally. I felt myself dealing with everyday stress situations in a more direct and assertive manner. I really couldn't believe the change in me. I didn't feel like a different person, but I did feel like an improved model of the old me.

In the beginning of March, I felt like I had hit a plateau and was no longer improving. I was still somewhat nervous, although much better than I used to be, and I was still having problems with words popping out of my mouth without first thinking about them. I found myself in a couple of embarrassing situations because of this. I discussed it with Dr. Fox and he suggested I increase my dosage to three per day, which should correct the difficulties I was having.

I was a little hesitant about the increase. After the side effects I had to endure to adjust to two per day, I didn't want to have to go through anything like that again if it wasn't absolutely necessary. He said I could try three and If I didn't care for the outcome, I could always go back to two per day.

I started taking three and was really surprised. The only negative side effect I experienced was a dryness of my eyes, which was easily corrected by eye drops every once in a while. As for positive side effects, there were many. I actually felt relaxed a good deal of the time. The word "relaxed" took on a new meaning in my mind. I felt as if I had never before in my life felt truly relaxed and comfortable with myself and other people.

When I was having high and low moods before I started taking the Lithium, I always assumed that "normal" was exactly halfway between the highest highs and the lowest lows. I expected the Lithium to get me to this "normal," but it hasn't. I find that what appears to be "normal" for me is a comfortable, moderate, high mood. I feel as though I have mentally lost weight. My head feels so much lighter without the stress of trying to cope with situations I couldn't correct anyway. I used to have constant headaches which were probably caused by self-induced stress. I very rarely have a headache now.

I just can't believe I'm having such a good time just being alive. When I wake up each morning, it's a new day and I can't wait to see what it brings. I still have an occasional day when I feel sort of blue, but not very often. When I do feel blue, it's not that devastating, lower than low feeling.

At first I was concerned about having to take Lithium for the rest of my life. The rest of my life is a very long time. Now that I've experienced what a difference taking Lithium makes in my life, I'm not afraid to take it indefinitely. I need to eat and drink every day to keep my body and mind working properly, so what's wrong with taking a mineral supplement that I need daily.

If I ever am foolish enough to stop taking it, I'm sure a brief run-in with the old me will be more than enough to persuade me to go back to taking it again. I like the new me too much to lose myself again.

Dr. Fox told me that he could make me well and happier in six months. He told me the truth. I feel well and happier than I can ever remember feeling before. I have been taking Lithium for six months, almost to the day, and I fully intend to keep taking it. I figure all I can do is keep getting better.

As you read through these patients' essays about how their lives have been affected by the mineral Lithium Carbonate, many questions come to mind. Obviously, there had been a selection process about which essays to present. However, because of my personal enthusiasm and considering myself an ambassador of Lithium Carbonate, I went out of my way to select those essays which contain a significant number of negative aspects or side effects rather than presenting the many essays which were simply overwhelmingly positive statements in support of Lithium Carbonate.

The preceding essay was written by Vicky, a thirty-four–year–old female. Her essay showed in graphic detail the extent to which a Lithium sensitive individual will go in order to be liked and/or accepted. The central theme of rejection, avoided at all costs, was rampant throughout this essay. As Vicky stated, "As early as I'm able to remember, I've been having mood swings." The hereditary, genetic, ongoing aspects of Lithium sensitivity are pointed out once again. An individual with a bipolar mood disorder, that is one who is Lithium sensitive, is actually no different at age thirty-five than at age five, only more so. What Vicky emphasized is that the difficulties an individual has with Lithium sensitivity and deficiency are ongoing and ever increasing. The increasing nature of the illness has to do with ongoing Lithium depletion. What I am saying is that a sensitive, moody adult was a sensitive, moody child.

Vicky also emphasized the importance of psychosomatic illness in a Lithium deficient individual. Lithium sensitive individuals speak through their viscera, through their physiologic states. Lithium deficient individuals have a great deal of difficulty verbalizing their needs. It was difficult for Vicky to say she was depressed and wanted to stay home. It was difficult and most often it is unrecognized. Consequently, her getting "sick" in a physical sense was a dramatization of her mood swings.

Vicky's essay pointed out very vividly and pointedly how much difficulty people have in obtaining a correct medical diagnosis of their problems. She clearly stated how she went from physician to physician along a wandering road seeking an understanding of her problems.

As her scenario unfolded and she began to discuss her father, it was obvious that he had a bipolar mood disorder, and that his moodiness and his methods of treating his family were rather compatible with, and almost pathognomonic of a Lithium deficient individual.

Vicky's interaction with her peers, in view of the fire, was an unfortuitous set of circumstances which cemented a Lithium sensitive individual's basic fears of insecurity and low self-esteem. Those comments enhanced withdrawal and isolation since criticism equals rejection.

Her interaction with her new group of friends clearly pointed out the chameleon qualities of Lithium deficient individuals. As she became older and her dating behavior became more intense, she once again had to adapt to the expectations of the people around her or fear their rejection. Her pregnancy was not unusual for such an individual.

The shame she felt about her pregnancy and the possible consequences of her relationship with her family were at best secondary to the overwhelming sense of fear, desperation, and loneliness when "abandoned" by David and Carol. Her ill-timed and basically unwanted marriage was part of the pattern in the early dating life of a Lithium sensitive female. As David and Vicky got to know each other more and more, they began to feel more and more rejection and withdrawal from each other. They recognized that neither really wanted to be married to the other and that their separation and divorce was a foregone conclusion. The only parameters which were uncertain at that point were where, when, and how.

The ongoing theme of a Lithium sensitive individual's need to please continued through most of the essay. Vicky was willing to please her parents, friends, doctors, nurses, hospital procedures, ministers, military officers, etc. She was ready, willing, and able to please anyone except herself. It is important to note that the intensity of the Lithium sensitivity was of such quality and quantity that she blended so with the needs of the people around her that she was truly not consciously aware of her own basic long-term needs and desires.

As you read how Vicky covered her son's face with a pillow, a light must have flashed in front of you. This was another clear example of how the overall concept of the "battered child syndrome" is directly related to Lithium deficiency and sensitivity. The inability to deal with stress and the need for impulsive, immediate gratification of needs and wants, are so evident in Lithium deficient individuals as well as parents involved in the battered child syndrome.

Vicky was a victim of her own naiveté, need for attention, and a hysterical type personality. The classic flirt is among the many characteristics of a hysterical type personality. Her lightheartedness and intense interpersonal affection is often misconstrued on a sensual-sexual basis. Lithium sensitive individuals are kissy, huggy, touchy, sensual people and are not necessarily sexual people all the time. However, the reputation as a "tease" can often cause serious difficulties for them.

The postpartum depression that Vicky suffered is a very frequent situation in which Lithium sensitive females find themselves. They have an insufficient amount of Lithium for themselves, but when the fetus extracts that last critical amount, depression results and often becomes intense. Another complication of postpartum depression is that by being depressed all of the time, they cause their loved ones to withdraw from them. It's depressing, negative, and undesirable just to be around them. Consequently, people who are around women who have postpartum depression may do anything and everything to avoid contact with them. This further compounds the problem.

Vicky's return to her home town pointed out how the Lithium deficient individual frequently starts many projects but finishes few.

I must once again comment on the extensive nature to which most Lithium sensitive individuals go to seek proper medical diagnosis and treatment. Vicky's case is not an overexaggeration of the great difficulty Lithium sensitive individuals have in finding out the truth. It is not for lack of trying, but because there is a lack of competent physicians who can readily recognize and treat the Lithium deficiency syndrome, better known as a bipolar mood disorder or a manic-depressive illness. The first psychiatrist to whom she referred was an obvious example of a therapist who has been trained extensively in one modality of treatment. It showed how such therapists will not be dissuaded from their appointed goals, regardless of the truths, facts, and information that are presented to them. They merely discard the information they don't want to use and amplify the information they do want to use. When they don't find the information they need, they often reroute the information they have been given to fit the needs they have to fit the system they have learned. What happens is that a trusting, caring, giving, loving, hurting Lithium deficient individual is misguided and misled. However, because of their own internal sense of desperation, they cling to these "healers" with a degree of tenacity and desperation which transcends any and all logic of the situation at hand. Unfortunately, the comments Vicky made about the psychiatrist who was no longer willing to see her because her insurance benefits ran out is also the rule of thumb rather than an exception. Unfortunately, her insurance ran out before she had the time to "get well." Had her psychiatrist seen her on a weekly basis, instead of twice weekly, she might have been able to see him long enough for the cyclical nature of her illness to take hold and for her to be able to get well on her own. Had this time frame occurred, her psychiatrist would have sat back and taken credit for her getting well. I do not mean this as a pun, but rather as a statement of fact: she ran out of cash; he would not offer her credit and she was dismissed from therapy.

Psychotherapy in general, and with me in particular, is a cash and carry operation. Psychotherapy, like any other field, must receive money in order to pay the bills and the employees. The difference is that with the proper recognition of a patient's illness and, in most cases, the judicious application of Lithium Carbonate, therapy can be reduced to a matter of several visits per season and, eventually, several visits per year, rather than several per week.

The varied and chaotic sexual and social lives of Lithium deficient individuals is more than correctly described and amplified in Vicky's essay. Periods of promiscuity, which parallel mood swings, are most common. These sexual interactions are often based not only on a need for an interpersonal relationship, but also are part of seeking out one's identity and an evolution of one's personality. The passive-dependent aspects of the obviously Lithium deficient female was described adequately.

I must comment that I found quite amusing the comments Vicky made about my first interactions with her, and then about our ongoing therapeutic interactions. I have often been described as harsh, pushy, and egocentric. I find these comments quite valid and essentially correct. My basic personality is that of a general practitioner. My personal experience with and knowledge of Lithium sensitive individuals is so extensive that I refuse to waste precious weeks, months, and perhaps years, on unnecessary discussions, while allowing my patients to continue to suffer. I am sure this approach has caused me to lose several patients over the years, however, because of its directness, openness, and sincerity, it has allowed me to treat hundreds upon hundreds of individuals who would otherwise be lost to routine psychiatric care. As you can see, my approach to a therapeutic problem is to evaluate it at its inception and to treat the here and now. To delve in depth into an individual's past is essentially unnecessary and even counterproductive in the overwhelming majority of psychotherapeutic patients. I am not saying classic psychoanalytic techniques are unnecessary. I'm saying that they are being applied to many patients not necessarily for the need of the patients, but because of the needs of the therapists. Psychoanalytic therapeutic techniques are invalid for an overwhelming majority of people to whom they are applied. Vicky is an individual to whom the use of these techniques was counterproductive and harmful.

I have learned over the years that you should never promise anybody anything. Vicky said I told her she would be well within six months. I doubt I said that, although it is possible. Rather, I'm sure I told her that she would in all likelihood feel a lot better after six months, and that within a year she would feel significantly better. However, at the same time, I tell my patients, and sometimes overstate, that although the time span might be extended for them to really feel better, the frequency of visits is less than one per month. Because the majority of treatment has to do with the medical model of replacing a mineral deficiency, the length of visits need only be fifteen or thirty minutes. This further reduces their time commitment, loss of time from work, and financial obligations. In addition, most individuals who are treated with Lithium spend well below ten dollars a month for their medication.

I recognize that not all therapeutic patients who come to a psychiatric practice need Lithium Carbonate. I further recognize that the overall number of people who come to therapy do need Lithium and that the vast majority of people who do need it go unrecognized by classically trained psychiatrists.

One aspect of my first interaction with Vicky, which seemed to upset her the most, was the ease with which I was able to analyze her personality. The reason was that along with the genetic transmission of the sensitivity to Lithium deficiency there are an overwhelming number of personality characteristics that can be readily recognized, defined, expended, and expounded upon. Vicky's fear and concern quickly became a therapeutic ally since she truly felt that I understood where she was coming from and therefore could help her, because my understanding of her personality was so accurate, my assessment of her treatment needs would in all likelihood be as accurate. I use this ploy frequently to help patients recognize their need for Lithium Carbonate and to be able to help them on the path toward mental health as quickly as possible.

WAYNE

You are reading these words because of Lithium Carbonate. Without it, I cannot slow my thoughts and emotions enough to communicate effectively. For years I went from doctor to doctor searching for help to stop or restrain my depression, sleeplessness, and often uncontrollable emotions. No solution was found that lasted more than a few months. Hospitalization, antidepressants, various strong sleep medications, different therapies, and even prescribed multivitamin treatments were tried. All missed the mark. The solution was simpler than these complex treatments. It lay in a mineral deficiency, a shortage of the element Lithium in my body and brain cells. There is an analogy to diabetes: a diabetic's body is short of insulin; my body needed Lithium. Another point in common is a lifelong need to continue medication in many cases. A lack of insulin can mean death to a diabetic; to me, life without Lithium meant a living death of mental instability.

The complexity of my symptoms and ongoing problems obscured the simple solution. My hands had been badly crushed in an accident at work. I had broken bones and injured nerves and the outcome of these injuries was a chronic pain syndrome known as causalgia. I have heard it described medically as one of the most profoundly painful syndromes known to man and I concur with that description. I was unable to use my hands in almost any way without suffering excruciating, disabling pain, but at first, I refused to give in to it. I worked at my telephone installation job in a limited capacity until I could no longer escape the danger I posed to myself and others. Driving became impossible. I lost my job and almost lost my family before I finally found a team of doctors who thought they understood my physiological problems and the mental and emotional trauma I had suffered as a result. Treatments and physical and mental therapy followed, test after test. The merry-go-round began to turn. It was a ride to nowhere that would consume more than three years of my life before I could finally step off, if only for a while.

At times, my moods soared to extremely unrealistic highs in which I acted impulsively, often to the detriment of my family and my self-interest. Thoughts rushed through my mind, flashing, barely discernible, like distant neon signs. Relatively short periods of exultation were followed by lengthy, deep periods of depression which bordered on suicide. My moods swung like a scythe through my life and reaped a harvest of despair. Perhaps that line is a bit dramatic, but its visual application is accurate. In those three years, I was hospitalized in three different psychiatric hospitals. I was treated by at least six psychiatrists who used different methods for various diagnoses.

As I mentioned earlier, a genuine, long-lasting, tolerable solution could not be found. Antidepressants had side effects which I could not tolerate. My mind was like a soggy sponge and all interest in life ceased while I took them. Some of these drugs had various sexual side effects and some caused intense weight gain. At times, within the context of the various treatment, they definitely served a purpose, but they were not a long-range solution by any means. Many types of sleep inducing medication were prescribed throughout these years, and they were used until they became ineffective. Sleep was my only escape from the listless, hollow existence my life had become. After a while, sleep became my primary goal as the ability to sleep longer than a couple of hours a night escaped me. Hospitalization became a major part of the process necessary to change the type of drug or decrease my medication. Throughout these years, no one possessed all of the information relevant to my situation. For some reason there was little exchange of data and recommendations between the different types of physicians treating me. My medical history quickly became expansive, contradictory, opinionated, and costly. Frustrated by failure and setback after setback, I refused to give up and be a physical and mental cripple for the rest of my life. My goal was reality, and the path was a treadmill of physicians and therapists. Progress was achieved and lost as I struggled to regain my sanity. A general description of my path to mental wellness through Lithium involves a summary of those struggles, as well as my experiences and hardships.

One of the first major obstacles to a return to normalcy was the fact that I had developed an addictive dependency to codeine, which was given to me on a prescription basis to control the pain of my causalgic nerve syndrome. After a period of addiciton lasting approximately eighteen months, my mind, body, and spirit began to repel this perpetual narcotic maintenance of pain control. "Chemical control of pain" was the neurologist's solution, and he felt it was the only hope available to me. I went cold turkey on my own and ended up in the University of Pennsylvania Psychiatric Hospital. They helped me by dealing to some degree with my emotional trauma, but codeine remained a part of pain control. A neurologist I saw there recommended an electric pain control unit that showed signs of helping others like me. It was there, in various types of physical and electronic therapies, coupled with psychotherapy, that I first experienced a pain-free period. It lasted all of twelve minutes, but it was bliss. The unit, a "Transcutaneous Nerve Stimulator," or TENS, helped me. I was able to cut back on the amount of codeine necessary to control the pain, and the first obstacle had at least been passed.

After a period of months however, the effectiveness of the TENS unit seemed to have lessened, and my codeine intake was once again increased. Within seven months, I had to be hospitalized again. For over two months I was in a clinic specifically designed to deal with problems of my type. I was admitted there because of my medically sanctioned addiction to codeine and my deep suicidal depression. Once again, there were varied therapies and modalities in treatment. I received help with the emotional side effects of my pain syndrome, and when I left there fifty-four days later, I had a stronghold on my drug addiction which over the next six months ended completely through a gradual reduction in intake. But the real problem, the genuine source of my difficulty, had yet to be discovered. I was seeing a psychiatrist weekly, but I was making little progress and within a couple of months after I left the clinic, I was once again extremely depressed and in terrible physical condition. My physical condition had deteriorated to the point where I required the services of an internist to help me deal with chronic duodenal ulcers. While I was under his care for an affliction that came about as a result of my condition, a solution was found for the root cause of my emotional and mental instability. While he did not present me with a cure, he recommended that I see Dr. Fox, a psychiatrist whose subspecialty was pain control through hypnosis and who was a recognized expert on cyclic mood disorders.

In only a few visits, Dr. Fox helped me control my pain largely through hypnosis. Then he began a holistic approach to my cure and another obstacle presented itself. After seeing so many physicians, I had lost trust and faith in the medical profession in general, and in psychiatrists specifically. It became Dr. Fox's job to overcome that lack of confidence and mistrust to the point where he could effect a cure, or at least a solution to my problem as it existed at that time. Within a year and a half, Dr. Fox had done his job so well that my wife and I decided to have another child. From shattered home, drug addiction, and institutionalization to a life filled with enough hope to want to bring another human being into it, but I am getting ahead of myself. That year and a half gave me very specific, practical insight into the actions and results of Lithium therapy. Some lessons came easy, apparent results and changes that enhanced life and made living worthwhile. Other lessons however were much harder to learn. Getting partially well after being desperately ill for so long was giddy stuff. Depression passed, I felt great, but problems developed once again. I found that I was the biggest obstacle to my being cured.

I was hospitalized for the initial application of Lithium and to terminate some of the other drugs I was taking. I continued to take Lithium for about seven months. My depression passed, I felt great, and then I stopped taking the Lithium capsules. My mood soared to high manic when I stopped. Within days I was riding so high that I felt there was nothing I couldn't do. I felt I was my old self for about two weeks, after which I was hit with a depression more profound than any I had known. I paid dearly for the first lesson: take your Lithium as prescribed for as long as it is prescribed. The return of normalcy in my life was a result of taking the Lithium, not cause to stop. The symptoms of manic-depressive illness returned like a catastrophic flood. I suffered through a unique hell of self-reproval and impatiently waited while the Lithium balance returned to my agonized brain and body. Once again, I waited for the merry-go-round to stop.

When I was again functioning in a progressive manner, I felt my problems were over but as I said earlier, after being abnormal for so long, I had to get used to a normal life. I had to learn to feel again, to allow myself to become angry in the course of events, and to cry and feel frustration without fear of uncontrollable emotions rising and destroying everything I had worked for. Over a period of months, my relationship with my wife and two sons, eight–and nine–years old changed from fear and mistrust to a closeness I thought we had lost through my illness and their suffering under my unrighteous hand. My inconsistent and unpredictable physical and verbal abuse as well as a total lack of discipline in my actions had alienated me from them almost completely. Friends had stopped coming to visit and there was little hope of restoring their faith in me. As far as they were concerned, I had gone too far too many times. I worked too hard to let disappointment and rejection stop me. Since there was no going back, I hung on. Lithium was the thread to guide me through the chaos of rebuilding my life. Make no mistake, I cannot overstate this: my life had to be completely rebuilt. Values had perished in the struggle to survive. My impulsiveness had shattered my self-confidence. My wife and I had taken separate bedrooms and our rediscovery of each other as I healed was a stumbling, fearful, but beautiful experience. Time passed and Lithium, as well as individual therapy sessions with a psychiatrist, caused my life to assume a comfortable degree of stability. I was almost myself again except for the fact that I could no longer work because of my physical injury. I was convinced at that time, as I am now, of Lithium's effects and the necessity of my taking it, but as time passed and conditions improved, I developed a system of rationalization which caused me to once again jump on the merry-go-round. Lesson number two is about to be taught: had I learned lesson number one properly, there would have been no need for a second one. My life situation continued progressively for the next few months, and despite all the warnings of my own experience, my insidious instinct for rationalization began to rear its ugly, subtle head. I once again convinced myself that Lithium was unnecessary. I told myself I had been cured and all the therapies were equally irrelevant. I began to take less and less and eventually stopped taking it altogether.

My life began to deteriorate in very subtle ways which I did not comprehend at first. I lied to my psychiatrist and my wife about continuing to take the medication, but with no idea of how to explain why I had stopped taking it. I rationalized that it made me sick and began to say I could not live life if I had to depend on taking anything every day, especially something which affected me in such a physical way. Finally, when I had just about destroyed all of what I had spent the last year and a half rebuilding, I was forced to admit to my doctor, my wife, and to myself that I certainly, absolutely required Lithium in order to control the symptoms of mental illness. So once again the rebuilding began.

In writing this summary, I have one primary intent: to relate the importance of maintaining a therapeutic level of medication as required in the treatment. Despite extensive fantasizing and rationalization, the need for Lithium in manic-depressive illness is inescapable. As my experiences in the hospitals have shown me, other cyclic mood disorders are also helped by Lithium. The fact that it works is unquestionable, but no medication, no matter how powerful or subtle in its effects, can work unless it is taken as prescribed.

I feel I have closed the door on mental and emotional illness and that it is locked with the heaviest of locks. Lithium is the key to keeping that door closed. I do not want my life to be what it was so I must continue taking Lithium for as long as I live, but it's a small price, a microscopically small price to pay for emotional, mental, and spiritual well being. I have heard much about miracle drugs, about useless, complex therapy, but the simplicity of Lithium and its lack of side effects in most people have made it to my mind, and to the minds of many of my fellow sufferers, a genuine wonder drug. I feel future generations will take Lithium for granted as a supplemental part of their dietary intake in one form or another. I am grateful to medical science and to the professionalism of the men who are fearless enough to apply it. That I am once again a whole human being proves its effectiveness and theirs. I thank all of you who have contributed to making me a useful, perceptive human being again, no longer deficient and finally allowed to work toward my fullest potential. I am grateful also for the opportunity to express my experiences and understanding of what I consider to be a modern miracle. I would like to close with the thought with which I began this statement, and that is, "You are reading these words because of Lithium Carbonate." The spectre of emotional illness is hard to shake, and without Lithium it would have been impossible for me. I had tried almost everything else. It is not just another treatment: it is the best, and my only hope.

As I read this essay, it became obvious to me that I have not fully elaborated on the fact that many manic-depressive individuals require long-term therapy. Although it is true that most Lithium sensitive individuals require short-term therapy with long-term followup, some individuals require intensive therapy. After we have made the diagnosis of a bipolar mood disorder, we are dealing with a category of illnesses. In the framework of Lithium sensitive individuals there is a vast array of personality patterns. Some individuals need and require ongoing, intensive therapy. Many individuals who are Lithium deficient have had significant trauma in their lives which needs exploration, understanding, and resolution. I want to clearly and effectively reiterate however that just because an individual has undergone a specific trauma in life that does not mean this trauma is responsible for the overall psychological state. Rather a single, severe emotional trauma could exacerbate an already sensitive situation or could accentuate it to the point where the underlying problem is then recognizable.

The extensive pain from which Wayne suffered was so overwhelming and destructive to himself and to his lifestyle that in addition to the judicious application of Lithium Carbonate, it also became necessary to treat his psychological problems. When we first met, Wayne's life was in a shambles. He was estranged from his family, his friends, and especially from himself. His treatment program required that all these areas be resolved, in addition to dealing with the most apparent problem of severe and overwhelming physical pain.

Wayne also pointed out many other important factors in the treatment of Lithium deficient individuals. It is a universal truth that as a whole men are not as cooperative as women in taking medication and keeping appointments. It is almost a rule of thumb that male patients will eventually stop taking their Lithium Carbonate capsules. This usually occurs within the first six months of treatment. Fortunately, most patients recognize the importance of Lithium Carbonate to maintain their lifestyles and quickly return to taking it, but many do not. They continue to watch their psychological states deteriorate and they do nothing about it. Men are at their best more difficult patients to deal with than women are. Wayne was no exception. Early in therapy, there was a great deal of verbal sparring and jockeying for position. Control is another issue which seems to be more prevalent with male than female patients. I am unsure why the control issue is an ever present theme. The patients are always in control. They take their medication, they come for their visits, they do one or the other or neither. Wayne was in control of his own temporary failure. He was in control of taking or not taking his medication. He opted not to take his Lithium and the results have already been noted.

It is noteworthy that one of the reasons he stopped taking his medication was a rather common situation with Lithium sensitive individuals early on in replacement therapy. They frequently report uncomfortable side effects from taking Lithium, like its effect on the memory and the use of the English language. Lithium takers frequently note that they have difficulty retrieving from their memory bank the exact words they want to use in sentences. They often find that for at least several months they use approximate words and words that get across the general meaning they want to convey, but not the exact meaning. Fortunately, with time, this clears up.

Many individuals are of such a genetic composition however that the problem with memory may be an extended one. Once individuals understand this they are usually able to overcome it.

Unfortunately for Wayne, he did not stick with the mineral long enough to allow this most common side effect to resolve itself. Consequently, he suffered.

Wayne is an unusually talented and diversified individual. Once he resumed taking Lithium Carbonate his talents became more pronounced and his ability to deal with life more consistent. He has been able to set goals for himself and stick to them. He has been able to set apart adequate time for his personal, emotional needs, as well as being able to devote ample time to his family. He is an exceptionally proud father and he spends a great deal of time interacting with his children. He is no longer the feared warlord but rather an individual with whom his family can relate and learn. Although the intensity of his suffering was great, the pleasure he now receives from stable mental health is even greater. He is a very happy and thankful individual whose newly found life has allowed him to live more comfortably and to enjoy the world around him more consistently.

One thing Wayne did not mention and which was significant with his clinical course is the fact that a great deal of time, effort, and energy is expended by the members of Lithium sensitive individuals' families while getting to know them again. Not only are the Lithium deficient patients at risk going through a significant number of readjustments as they relearn their own personalities, but the family readjusts on an equally intense basis. In Wayne's case, his wife's adjustment period was so overwhelming that she required therapy. She was not sure about whether or not these positive changes in her husband's personality would be ongoing or as shortlived and as unstable as with previous treatment modalities. Although life with him was erratic and painful, she began to become accustomed to his inconsistencies.

Until Wayne came to therapy he was "consistently inconsistent." Although he was difficult to deal with, his family had become accustomed to his outrageous and demanding personality. They were not quite sure how to "take" the new him. They were not sure whether these changes were to be adapted to or whether they would be fleeting.

As with most modalities of therapy, especially in the area of psychiatry, but also in general medicine, trial and error in applying a specific therapeutic modality is commonplace. As my experience with the use of Lithium Carbonate grows, the necessity for involving the family in educational and instructional sessions has become an ever increasing part of the overall therapeutic approach to Lithium deficient individuals. It has become commonplace for a family to be included in these educational and orientation sessions so they are intellectually prepared to deal with the patient as he continues to change and grow. As we have already mentioned, Lithium sensitive individuals are dependent on and extract emotional support from their environment. They are individuals whose locus of strength comes from without. Since this is an integral part of the personality of Lithium deficient individuals, the external support system must be properly educated and informed about their role and what is expected of them during the therapeutic process.

When a story is told or instructions are given, irrespective of the subject, there is great variation in how the information is received, and an even greater variation in how the information is retransmitted to others. Consequently, it has been my practice never to talk to any member of a patient's family unless the patient is present. There are some exceptions, especially when dealing with patients who are extremely ill, but as a rule of thumb I adhere to family sessions as a common practice. During the initial phases of explanation, Lithium sensitive individuals need a great deal of encouragement, and often this can be provided by knowledgeable family members. We also ask family members to relate their observations of the patient's behavior so we can have a clear picture of what is happening. I do not want the patients to bring me "gifts"; I want them to tell me what is actually transpiring and having family members present augments this process.

Wayne went through a great deal of emotional pain and anguish while learning to understand himself and his orientation to life. As his emotional pain decreased, he was better able to handle his physical problems. Depression and anxiety cause true physical pain to be amplified even more. It is our practice to approach pain in a multifaceted manner. The emotional aspects of the pain syndrome are treated at the same time the pain syndrome is treated. The results are more satisfying and more rapid when patients are handled in a multipronged manner. Wayne and his family are currently closer and more integrated as a unit than ever before. His desire to get well and the support of his family have brought him out of the depths of depression and back into the mainstream of life.

THE EMERSON FAMILY

The following patient presentations were selected to demonstrate the ease of diagnosis of manic-depressive illness, as well as a longitudinal family history. This example clearly delineates the ease with which the diagnosis can be made if the physician is willing to openly evaluate and examine the data as a patient presents it.

ELLEN

Over twenty years ago, I was a young, warmhearted college student who became intrigued with a moody, intelligent, and mysterious fellow student. During the months we dated, he grew increasingly dependent on me and his need for me made me feel very important. Even though I had had many boy friends and was deeply in love before this relationship, I had not felt so needed. I was therefore intrigued and flattered.

While we were still dating and in college, Steve would drift into periods of gloom and despair. His whole expression and posture would reflect how he felt. This was a challenge to me as a bright-eyed twenty-year old. I felt satisfaction bringing some relief at his times of despair, which usually had no apparent cause. At other times Steve would be so full of himself he would talk at great length (in a monologue) about how he wanted to change society. He conveyed a general feeling of superiority to other mortals. It was as if he was a different person at different times.

There were problems for me in this relationship even before

we got married. Steve was extremely possessive and eventually showed a strong temper especially when my parents acted concerned about my involvement with him. He would say, "If I can't have you, no one can," and act desperate at the thought of not having me. He introduced the element of fear and that remained in the relationship to the end. I confused Steve's great need for me with what he claimed was love, so I married him. (He showed over the years an inability to give, to love, to care for anyone but himself. Other people were important only in terms of what they could do for him).

The marriage was terrible from the beginning. I was raised to "stick with it," and since I was a strong person, I believed I could help Steve and make a decent life out of a mess. Unfortunately, I knew nothing about manic-depression or other forms of mental illness. When Steve said my family was crazy, I wondered if that wasn't true, for he was wearing me down, bit by bit. I thought his family was strange but I was afraid to say so.

As part of his recurring grand scheme to save society, Steve decided to go to a seminary in Princeton. I was the good wife and by now a mother (which really made me feel stuck in the marriage), so I did all I could to help which included a return to full-time teaching. He did all right academically because of his level of intelligence and good use of words, but at times he wondered if he wanted to continue. He did not really believe much of the theology. In fact, he never liked attending church unless he put on the show himself.

One summer, we stayed in Nashville, Tennessee with my aunt and uncle while Steve worked with their presbytery. I was told years later that the presbytery was concerned about Steve's emotional stability. They thought they were doing me a favor by not telling me that at the time. The responsible observations of such people could have been a tremendous help to me.

After he completed the seminary work, graduated and was ordained in the Presbyterian Church, Steve wanted to go into inner city work and be a savior in the ghetto. I told him I would not do that or consider taking our young son. Steve had always expected me to be supportive of whatever idea he had. This was one of the few times I stuck to my own feelings despite his reaction. He reminded me periodically over the years how I kept him from what he wanted. When I finally would answer, "Go do it!" he could not.

We spent the next five years in Dallas, Texas where Steve was the minister in a church. He complained about the church, session meetings, little old ladies, raising money, and being a pastor. The one thing he liked to do was speak to groups, the larger the better. I despised the phoniness of it. When he was preaching, his voice would take on a special tone. Even our young children recognized it as "daddy's preaching voice."

Steve's ideas about what he wanted were always changing. There were some themes that were repeated, but they could not be fit together as a whole. When he decided to return to the seminary for a year's graduate work, I went along in order to get out of the church mess and return East.

Many of the long-term, underlying problems continued, and some intensified while we were back at Princeton. Steve would change personalities so quickly and usually for no apparent reason I began to wonder if he was schizophrenic, for I know he was not the same person for any predictable length of time. I could see no real cause for these frequent personality changes, nor could I predict the time he would change. I just knew the topsy-turvy aspect was always there, no matter what superhuman efforts I made to stabilize things and to keep the situation steady. It was obviously more than I could handle.

Steve became involved, actually absorbed in the civil rights movement and was arrested in Camden, New Jersey for civil disobedience. (A few years later he had the conviction "expunged" from his record.) He enjoyed the activities, felt important as a leader, and seemed to be carried on a wave.

Although he remained a Presbyterian minister, Steve never returned to parish work. After a short time, he would lose interest in every job. In the space of a few years he worked in the presbytery, taught in a Friends school, sold life insurance, and worked for the New Jersey Division of Youth and Family Services.

Earlier I referred to Steve's "normal" family, in contrast to what he called my "crazy" family. At this point, his family showed it had serious and tragic problems. His mother committed suicide by hanging herself. His brother and sister were hospitalized (at separate times) in mental hospitals, both with severe depression. His sister had hallucinated and attempted suicide by cutting her wrists. She was given a series of shock treatments.

Even though I would have liked to have been removed from the whole scene, I tried to show human concern for Steve during all these crises in his family. It was difficult to reach him. He was frequently in a frenzy. He finally saw a psychiatrist, but it was not helpful. Steve was well read and articulate so he could talk circles around the therapist. Nothing was ever said about manic-depression nor was any connection made with the problems of other members of his family.

The moods continued to fluctuate and sometimes Steve expressed rage and temper that were frightening. The children and I learned to tiptoe around him so we wouldn't set him off. Meanwhile, most of the outside world saw a very different person, a nice, respectable man. When I read news stories about "normal, respectable people" who did horrible things to their families and shocked everyone, I had a very uncomfortable feeling of seeing how things like that happen.

It was finally brought to my attention that Steve sounded like a manic-depressive and may have had a chemical imbalance that caused much of his erratic behavior. He was somewhat interested when I talked to him about this idea especially since he was in a depressed state at the time. He went to the clinic at the University of Pennsylvania Hospital. With his symptoms and family history the doctor considered him a definite candidate for Lithium Carbonate therapy.

Even though I felt our life and relationship was hopeless, I was encouraged that Steve might finally be getting medical help for his very serious problem. Unfortunately, he never received the full benefits of Lithium because he took it in a haphazard way, not on a regular basis as prescribed. Sometimes he took a little too much, got shaky or nauseous, and then wouldn't take it for a while. This was typical of his seesaw behavior. With Lithium therapy, a proper Lithium level must be carefully maintained everyday, so this haphazard attitude only added to his bleak, personal situation.

Steve seemed shocked and devastated when I divorced him on grounds of extreme cruelty, but I knew that when he would not give the Lithium a serious try he did not want to be helped, would continue to hurt himself and others, and I would no longer be one of the others.

The latest report I have is that Steve has not taken Lithium for months, has quit his job and has gone to Los Angeles to "find himself" at age forty-five.

I have discussed some specific aspects of living in a marriage with a manic-depressive husband. I would now like to make some personal observations about raising two manic-depressive children, Mark and Sharon.

Mark, our first child was a healthy, well developed, and very demanding baby. He had a ravenous appetite for food and attention. Mark was a very social creature. He communicated early and talked in simple sentences by age one. He was physically mature also. He discovered how to do somersaults at eighteen months and taught himself to ride a friend's two wheeler bicycle at age four. (He insisted on no help.)

As a mother and teacher, I knew Mark was no ordinary child. I felt he was both bright and physically adept. Unfortunately, I knew nothing of manic-depression or Lithium deficiency until he was twenty. Many of the upsets and problems that appeared over the years would have been greatly alleviated if we had known about Lithium therapy.

The first area of obvious distress was his underachievement in school. His teachers were constantly saying, "We know Mark's bright (his I. Q. in elementary school was 135), but, we don't know why he won't put out any effort." His teachers also observed him daydreaming, "as if his mind is off somewhere else." He showed every sign of being turned off by school and he hated books. He even refused to read directions to put something together and actually assembled an entire bicycle just by looking at the parts.

In high school he was quite popular. He had the lead in the senior class play and was chosen for the Honor Society. He claimed he got through much of his course work by "shooting the bull."

There was a crisis during his junior year when his nose was badly broken during a basketball scrimmage. It was necessary for a plastic surgeon to repair his nose and sinuses. The doctor insisted on a two-week wait before he could return to basketball practice. Since he was not able to go back to practice until just before Christmas vacation, the coach announced to my son (the tallest boy in the school) that he had filled the opening on the team while he was away and had no place for him.

I remember that when he came home after getting that news he was so upset that he went directly to his room, alone, and he cried. For the entire vacation he was desolate and extremely depressed, and then he became ill with bronchitis.

After high school, Mark entered an architectural school at a university hundreds of miles from home. The combination of the new environment, stress from the academic pressures (after never having worked in high school except at varsity sports), and trying to handle daily living responsibilities added up to too much of a load and during the second year he dropped out.

The strain at home was getting increasingly worse, especially between father and son. After a series of unsuccessful jobs Mark decided to go to California, where he stayed for two years. He had an assortment of jobs and lived in a series of apartments with other young people. It was a case of barely surviving, both in the physical and emotional sense. He had a bad motorcycle accident, used drugs and alcohol, and had many emotional ups and downs. He described times when he would become hyper and unable to slow down, and conversely, times of discouragement and despair.

Mark's return to the east coast was as sudden as his departure for California had been. Since I had become convinced the problems were typical of what I had learned of manic-depressive illness, and because of our family history of symptoms I talked with my son, who was quite receptive and willing to get help. He said he felt relieved of the "hyper" feeling within days of starting the Lithium therapy.

Over the following months there was steady improvement in his emotional state, with a few setbacks. The worst episode happened within a few weeks of starting the Lithium therapy. He drank so much beer and bourbon that he blacked out and could not remember the events that followed.

Life has become increasingly stable for him although his basic funloving, gregarious personality has remained the same. He is doing well in his job and has recently married. He is extremely careful to take his Lithium regularly. He takes no drugs and is cautious about alcohol. He's able to joke about his need for Lithium, but he also knows it is a very serious matter and is thankful to have its benefits.

Our daughter, Sharon, came into the world with a wide-eyed look on her face. She was a calmer baby than her brother, but she also had a strong need to be cuddled and to be with people. She was quite responsive to any verbal or physical contact. She has always had a never ending need to talk. It was humorous to listen to her "talk" constantly as a baby. Even though she did not know any words, she "talked" in very expressive baby phrases.

All of her life, Sharon has been supersensitive and needed only a look to detect disapproval. This trait grew into a problem for her, as she constantly feared rejection or scorn from others. Unfortunately it overshadowed her many real talents and abilities.

She was bright enough to be accelerated from sixth to eighth grade and to go through high school in three years, yet she lacked confidence in herself. She was extremely creative, could make something out of nothing, continued to have a wide-eyed curiosity, and became a good dancer. Still her self-esteem was low. I listened to her try to talk through her worries and anxieties for hours at a time. There was one long period where every afternoon after school she would either talk to me for a long time or, if I was not available, she would go into a deep sleep. I was too close to the situation, and because of my horrible marriage, I did not realize the depths of her depression. Even if I had realized it, there would have been few doctors aware of what to do to help her.

It was very difficult for Sharon to take any criticism, particularly from her ballet teacher. She would feel she was an impossible mess and that the teacher did not like her. She also thought of herself as a very unattractive person. It came as quite a surprise to her as she matured that she received compliments on her appearance.

Even though Sharon had excellent grades in school, she did not establish good relationships with many of her teachers. She could be critical and difficult (and talked a lot, of course).

At times, she had an almost inflated ego or vision of things she could do and she overloaded herself, and at other times she would be completely down and withdraw into herself. It was difficult for her to make a fair evaluation that did not go to either extreme. This trait, or problem, has been greatly modified by Lithium therapy.

Since Sharon was barely sixteen when she graduated from high school, we thought it would be best to wait a year before she started college. That year turned out to be a disaster for her. She spent many lonely days and was too involved in the strains of our marriage, which was about to terminate. She developed an extreme case of anxiety. We were fortunate to be able to get her excellent psychiatric help and soon realized that she had problems from her whole home environment as well as the inherited manic-depression. Therapy has been tremendously helpful. When she started Lithium therapy, there was a very noticeable difference. At age sixteen, she felt she had reached "rock bottom." After starting therapy the positive changes in her life were fantastic.

When she started college she was still taking her Lithium and sometimes also Vivactil for depression. The depression has lessened greatly and she's able to tell when she's getting manic and can try to curb it. She has frequent blood tests to keep the Lithium level just right. I'm convinced if she had not gotten the proper therapy and Lithium, she would not have been able to successfully handle these first two years at college. She now plans on spending her junior year in France.

Sharon is still wide-eyed and curious, lovable, talkative (but also able to listen now), and she has a healthier opinion of herself and life now. She talks quite openly to students and professors about her manic-depression and Lithium deficiency. It's almost as if she has found such great help that she wants to share it with other people who may still be suffering.

Ellen was sent to me for treatment of an ongoing, severe depressive syndrome. She was referred to me by her family physician. Fortunately, he is a psychologically minded and knowledgeable individual who also uses Lithium Carbonate in his practice. As Ellen began to unfold her history, it became obvious that her husband, Steve, had a bipolar mood disorder, circular type. It further became obvious that I could not treat Ellen without also trying to effect proper medical psychiatric care for her husband. Consequently, after I explained to Ellen what was wrong with her husband and what needed to be done, I suggested he seek treatment at a major center. I elected not to involve myself in his therapy while treating his wife. As her scenario unfolded, and as she spoke with significant desperation and severe emotional pain, the strong possibility of divorce was ever present. I felt that to treat him and to treat her at the same time would be too much of a division of allegiance.

As her history began to encompass that of her children, I was more than willing to see both her son and her daughter and to help them stabilize their lives and establish a normal Lithium level. As in any field, the rules and regulations which psychiatrists follow at any given point in time reflect the mood and state of the art in that specific area. In bygone days, it was considered heresy to see more than one member of a family at any time, but this is absurd. Being able to accumulate a history for various members of a family simultaneously allows a physician to significantly decrease the number of visits and the overall time for treatment and markedly decrease the expense of treatment. Because of the complexity of Ellen's lifestyle, she was seen on a weekly basis in active psychotherapy. Sharon was being seen in active psychotherapy as well as Lithium Carbonate therapy.

Although it was rather easy for Ellen to recognize the manic-depressive qualities her husband exhibited, it was much more difficult for her to deal with the fact that she had a Lithium deficiency, too.

Lithium sensitivity can present itself with a wide variety of symptoms. Ellen's sensitivity was primarily in the area of low self-esteem and overwhelming sensitivity to rejection and criticism. She was an extremely emotional individual who became easily depressed and was frequently overwhelmed by her environment. She was very emotional which caused her to frequently subjugate her own needs to those of the people around her. She was easily programmed by her husband and her self-confidence eroded. She was an individual whose need for Lithium Carbonate was very subtle, however, because of the benefits she had seen in her children, she was willing to try it.

One of the many facets which Ellen's history revealed is the tenacity with which people will stay with manic-depressive individuals because when people with manic-depressive illness are well, they are usually nice to be around at least in the early stages. Additionally, they instill within their mates the need to be needed. This becomes very painful, but early in Ellen's situation it was rather rewarding and refreshing. Many of us are willing to "stick it out," as Ellen stated she was willing to do. A crisis often occurs in this philosophy when an answer to a problem has been decided upon or elucidated, and yet the individual is unwilling to cooperate, such as Steve's haphazard attitude toward his Lithium therapy.

It was Steve's unwillingness to cooperate with his Lithium Carbonate therapy that finally made his wife realize the futility of the situation. She did remain with him for well over a year after his introduction to Lithium Carbonate before finalizing her decision.

At the expense of being repetitive, it is important to once again emphasize the fact that manic-depressive illness is becoming more and more prevalent with each generation. I am sure you can appreciate the frustration Ellen suffered having to deal with a Lithium sensitive husband and also with two Lithium sensitive children. The moodiness, impulsiveness, intensity of the interaction, but most of all the inconsistencies in dealing with her loved ones finally wore her down.

The Emerson family illuminates the point that many individuals who are Lithium deficient also need a therapeutic environment in which to grow. It must be remembered that individuals who are Lithium deficient are dramatically subject to the emotional environment around them. Consequently, changes in the external emotional environment must be made concurrently with any changes the individuals make in their internal emotional environment. In this situation, when the external emotional environment (her husband) was unwilling to take measures to help stabilize his moods, Ellen was forced to make a significant change and obtain a separation and divorce.

Individuals who are Lithium deficient are often very intense and very convincing. It is not unusual for interactions like Ellen and Steve's to create an atmosphere in which the healthier individual is convinced that he or she is ill. When I first met Ellen, she was convinced that her problems were of her own doing and that she was in need of extensive help.

As we analyzed and investigated her life, she recognized that she did not necessarily initiate many of her problems, but that her's were errors of omission. She would not or could not separate from her painful marriage for many, many years.

Errors of commission are those which are made by word or deed. Errors of omission, although just as troublesome, are made by not saying or doing something. At different times in our lives, we are all guilty of both, however, errors of omission are often more subtle and more difficult to recognize. Ellen's emotional collapse was caused by errors of omission, such as allowing Steve to dominate her life. Once she understood this, she was able to make practical changes in her life.

SHARON

Lithium and my psychiatrist saved my life. I cannot say that Lithium alone put me back on my feet and gave me happiness; I had a lot of help from my doctor too. I must admit that without Lithium there would be a lot more unhappiness left for me to endure, most likely for the rest of my life.

Unfortunately, I was born a manic-depressive in a family full of manic-depressives (which is not all that uncommon), with a manic-depressive father who had enough psychological problems to distort any child's mental health, manic-depressive or not. I needed a doctor to deal with his effect on me; I needed Lithium to deal with myself.

At times it's difficult to differentiate between the help I've gotten from my doctor and what I've gotten from Lithium. I think the simplest solution is to say that having been born with a Lithium deficiency, I have certain innate tendencies (moodiness, fear of rejection, constant need for approval) and that having a psychotic, manic-depressive father, all these traits were intensified and become much more detrimental. If I had been born into a loving, accepting environment without the additional negative stress, I could perhaps have sublimi-nated my manic characteristics and become happy, although eccentric. Frankly, I believe I prefer taking Lithium to living the life of a manic without it, even if the extra stress wasn't threatening my stability.

I started out as a typical manic, but withstood about sev-enteen-and-a-half years of constant, relentless put-downs; nothing I did was worth a cent and I didn't matter much as a person myself. To make things worse, I was manipulated to feel guilt and confusion by being told (not shown) that I was indeed loved. All this was somehow to satisfy my father's needs and since I had never lived without it the fact never occurred to me that my situation, like my mother's and brother's, was abnormal or that I was being wronged. I simply grew to believe that I was worthless and consequently I expected the worst to come from everything. With the self-esteem of a worm, I messed up any relationship I managed to have. I was always building up expectations which by their very nature had to let me down. I always had the feeling that everybody disliked me. It's a wonder that anybody liked me at all considering the way I felt about myself.

Speaking of self-esteem brings up another one of my Lithium deficient characteristics: I lived for what others thought of me. I had to have approval. I did anything I could to get attention, even if it was negative. My doctor says negative attention can seem better than none at all. I had to find ways of being made over, sometimes by shocking people, by showing off, or by "brown nosing." I was always doing everything possible to win the attention of boys, but I lost interest once I knew they liked me. Because I wanted so much to be loved, being put down constantly aggravated my problems. People only amounted to what they provided me in terms of satisfying my basic needs. Constant rejection made my genetically exaggerated fear of rejection most intense. Since all manics take criticism personally, I took all criticism as rejection.

I was programmed to overreact to *everything*. All letdowns devastated me. I internalized all my frustrations and became depressed. Sometimes my anger would surface and I'd let it out on the nearest victim. This would always get out of control and the initial point that stimulated my outburst would be lost. I could end up fighting about anything as long as I came out on top. I had to be right, no matter what, and I would practically strangle someone to win. During the same dispute, I could completely switch my argument and defend the opposite side if it helped me not to be proven wrong.

My fear of rejection showed up most clearly when I began to get proficient at music and dancing. I felt I was basically untalented though I practiced very hard and I was dreadfully afraid of making mistakes and being rejected. The better I got, the worse I feared rejection, especially at an audition where I felt more was expected of me. I developed a paralyzing performance anxiety that struck every time the opinion of my audience mattered or seemed likely to be critical. Acutally this was a subconscious trick of my own so that I could always fail because I was nervous. Even though I was embarrassed, I could blame it on nerves and not on myself or my inability to perform.

My mood swings were always severe, but they finally interfered radically with my life when I became a teenager. My depressions would be set off by the most minute disappointments and then snowball to a point way beyond the issue that bothered me. Once I was depressed, I had to either wait for some lucky event to perk me up, jog ten miles, or simply wait until the cycle brought me back afloat. Usually, I didn't even want to feel better. That is scary because I walled myself in until I felt so alone and black that I could not function. My manic highs weren't all that rosy either, though many manic-depressives are defensive of their highs. When I was manic my judgment went down the drain and I acted as if I were on speed. Inevitably, I crashed every time. I used to think that being manic meant being happy since I had never known what it was to be truly happy and have peace of mind.

After I acknowledged that I could not stand myself because someone else had convinced me not to, I began to learn to love myself. Most of my subsequent happiness was achieved only after beginning Lithium treatment. Within a week after I started, I found that I was calmer, but not drugged. I began to stop reacting to things that used to enrage me. Over about a year's time, I felt a gradual change come over my personality. My depressions steadily grew further apart, were shorter in duration and less intense. Recently, I found myself feeling somewhat like I used to when I was slipping downwards, and I got rather scared. I realized that was bothering me and was determined to fight it, even though I was sure the depression would last a long time. To my great relief, it subsided in one day. A year ago, I would have been just short of desolate for weeks or even months.

After taking Lithium for a while, I noticed I wasn't jumping down people's throats if they upset me. I was holding my tongue much more at times when I once would have been verbally critical. I was keeping more friends than I had been, and a sense of peacefulness came which I had never had before.

There is one thing that stands out when I think of how different I feel: for the first time in my life I know how to love people for that they are. I was always loving, but I had so many needs as a manic that most of my efforts had to be centered on me. I would sort of fall in love, but I usually did so with someone who could give me the approval I needed. I suppose I'm just talking about infatuation, but it's the only love I knew. I built up images around people, but of course they could not live up to expectations they were not aware of, so I felt hurt. I even extended this to my mother. My way of loving her was to love what she did for me. I could not spare the emotional energy on truly caring about her. My concept of romantic love is so different than before that I must say I have never really been in love; I didn't even know what it was. Even now, after almost two years of Lithium therapy, I feel certain that I will learn more and more as I continue taking it.

Without a doubt, I am not the same person I once knew. I was moody, volatile, lonely, nasty, and miserable. Now, I love myself, I'm loving, peaceful, accepting, productive, I'm growing, and I'm happy. I understand people around me and I feel like listening to them. I'm even grateful for things I never appreciated before even though I had been so dependent on them. I am not a slave to my emotions; I can wait for things and even enjoy myself while I wait. I actually think I feel more, too, and I don't intellectualize or try to escape what I feel.

I will take Lithium every day for the rest of my life, even should they tell me some day that it causes cancer. (By the way, you can't skip pills or you have to start all over again, believe me!) Once you have something as wonderful as Lithium, reverting to being a maniacal manic is just about the worst thing I can think of. I would give up almost anything before I'd give up Lithium.

Sharon's statements at first appear melodramatic and overstated. The best comment I can make is none at all. To have known Sharon before Lithium and after taking Lithium would more than corroborate and substantiate her statements. To add anything else would at best be redundant.

Mark's therapy has consisted of a fifteen-to-thirty minute visit once a season to renew his Lithium prescription and test his blood. He had neither the interest nor the motivation for more intensive therapy. Fortunately, most of his difficulties have come under his control with the use of Lithium Carbonate alone. His response has been quite significant and rewarding.

The Emerson family history demonstrates the variety of therapeutic approaches one can use when dealing with Lithium sensitive people. The cornerstone of therapy consists of the use of Lithium Carbonate. Lithium Carbonate therapy alone is often enough, but not always.

Delineating the Emerson family tree underlines the importance of obtaining a careful history of the mental health of as many members in as many generations as possible. One of the most useful diagnostic criteria for establishing an individual's need for Lithium mineral replacement is a family history indicative of manic-depressive illness.

HARRIET

My decision to begin using Lithium was not an easy one. I was filled with much fear, anxiety, and apprehension. What was Lithium? Exactly what would this element (I had been told it was not a drug) do to me, both now and later in life? The knowledge that many medications left individuals with undesirable side effects made me even more tense about my decision.

My doctor had given me a small pamphlet about Lithium, but it did not answer many of my questions. It briefly discussed Lithium, its use for manic-depressives and for various categories of manics. While I could acknowledge a few aspects of my personality in these categories, I did not think I truly fit into any of them. (I also felt that a mentally healthy person could find a part of himself or herself in those categories.) I felt accepting the need for Lithium labeled me a manic-depressive.

I had no qualms about admitting that I had a mental health problem, but I did strongly object to being labeled.

The fact remained that I was told I needed to take Lithium and if I truly wanted to get better, I had to make a decision about whether or not to accept using this element.

I decided I needed more information. I did not discuss this with my doctor since I did not want him to feel I was questioning his competence, which I was not. I was, in fact, apprehensive and scared, scared of a drug (while I knew it was an element, my mind continued to consider it a drug) that I knew nothing about and about which I could not easily find any information. I set about using many of the resources I had available. I spent time in the library, but found very little information. I talked to friends who were nurses and who had worked in the mental health field. No one knew any more than I did. I even went so far as to have a friend who worked at a hospital check with a psychiatrist she knew. The results were basically the same: it was an element that people frequently had a deficiency in, it was compatable with anything else, it helped alleviate depression, and no one knew exactly how or why it worked. He also said it was used not only for manics, but also for those suffering from depression and for some children with behavior problems. This information was interesting, but I was still left with many questions. I had to make a decision.

I was still quite apprehensive, but I decided to use the Lithium. Things were going rather smoothly in my life at that time. I guess you could say I was on a temporary natural high. I was not willing to consciously or subconsciously admit to the fact that the Lithium was having any effect on me at all. I believed I had everything under control. Therefore, I would sometimes forget to take the medication. I still questioned it and the need for it, especially since I hadn't felt depressed in such a long time (a whole month).

When I was told my five-year-old son was suffering from a Lithium deficiency, that compounded my feelings about using it. If I decided I was not going to use Lithium, then in good conscience I could not give it to him either. Since I still had a lot of unanswered questions about the side effects Lithium would have on me, I could not help but wonder about any possible side effects it would have on him, especially at his young age. If I decided it was necessary for him to use the Lithium and that it would not harm him, then how could I intelligently justify my own refusal to take it, other than a matter of ego.

I never had such a difficult time making a decision. Every time I took this little pill, I questioned myself. Why? Was it necessary? I felt so much anxiety that I can't help but wonder if possibly the advantages of the Lithium were not hindered. I had not personally experienced any side effects but I did not accept that. I had a mental block about Lithium, yet I was not willing to give it up. In a small way, it was a form of security. The unknown was scary. If I permitted myself to objectively look at Lithium and its benefits, I may have had to admit it was needed. Needing it, using it, and accepting its benefits, would change me and my perspective of things and thus take away all my security. I was secure with my depression, my feeling of inferiority, and the sense of being unliked. How could I give all this up? What if I didn't like the person who really existed? To change my entire viewpoint and how I perceived my life was a very uncomfortable thought. I had always wanted to be liked, but I never felt I was. Now I was being told that taking this element would help me have a positive self-image, something I wanted so badly yet was running away from. I found everything to be very confusing.

I had been taking the Lithium for about a month and a half when I first decided to discontinue it. The thought had been going through my mind for quite some time. It was just after Christmas and I had the flu. I decided that would be a good excuse to stop. I spent a lot of time rationalizing the reasons for discontinuing it. I even felt I had a good excuse to give my doctor when I told him I had stopped. I truly believed I would tell him I felt great, I was handling all my problems without any difficulty, and I no longer needed his Lithium. (I thought of the Lithium as his, something he just came up with.) Things didn't go quite as I had planned. My irritability and temper weren't as bad as they had been, but this was probably partially because I was sick in bed. I attributed the irritability I was feeling to the fact that I was expecting my period. In my mind I had completely justified not needing the Lithium, yet, something was nagging at me and I could not pinpoint what it was.

When I went to the doctor several days later, I told him I had stopped taking the Lithium. He asked me why and I told only half the truth, that I was sick. We discussed it for a while and then he asked me what I was going to do. I told him if he felt I still needed it, I would start taking it again. I should have expected his reply. He told me to go back on it and stay on it faithfully, that for it to work properly, I had to take it properly. I said I would. That lasted two weeks at the most.

Friends had told me that you could not drink or smoke pot while taking Lithium. This upset me immensely, but not for any really good reason. I did drink once in a while, but I hadn't smoked pot in my life and at thirty-two, I probably wouldn't, but it was a good excuse to stop the Lithium. I decided my doctor hadn't been completely honest with me. If he hadn't told me about this (I did not bother to question him about it), then what else hadn't he told me about Lithium? That was all I needed: no more Lithium for me. Oddly enough, I kept my son on his medication throughout all of this.

I did not acknowledge any of the changes that were going on within me, or the fact that it was taking almost all of my energy to keep my composure. My personality was changing drastically. I would cry at the drop of a pin. My kids were constantly getting on my nerves and I was continually yelling at them. I felt as if everyone was talking about me. It was seldom that I was not in a state of depression. I was a complete bundle of nerves. Even though my friends were concerned and repeatedly asked me to go back on the Lithium, I refused. I knew I had no need for this element and would be able to control things myself, given time.

After two weeks, I was beginning to question my decision. I was still depressed and irritable, but now I was beginning to react physically with my kids and that scared me. Somehow I had to get my questions about Lithium answered and make a decision I could live by. Since I had an appointment with my family physician that Saturday, I decided I would discuss it with him.

I wanted my questions answered, but I did not want him to know I was on Lithium (or supposed to be on it), nor did I want to discredit my doctor. I decided the best way to handle this was to say I was doing a school paper on Lithium and its use and to ask if he had any information on it since I was having a difficult time obtaining any. He was very cooperative. He told me he used Lithium for patients who were depressed and that it required monitoring by blood test (which I already knew). When I asked him about using alcohol and pot while taking Lithium, he told me patients were told not to use them because they were not good for them, but that they caused no side effects relative to the Lithium. We discussed the fact that no one really knew exactly how it worked. He also lent me his *Physicians' Desk Reference* so I could obtain additional information for my report. I was relieved; he hadn't suspected the truth.

While my excuse for discontinuing the Lithium no longer existed, I did not resume taking the medication. Once again, I had to make a decision. I had an appointment that Tuesday with my doctor and I knew this was to be a final decision. I was quite perplexed.

On Sunday, I tried to spend a lot of time thinking. My irritability was at a peak and my kids were literally driving me up a wall, especially my five-year-old son. Suddenly, before I knew what was going on, I was screaming and yelling at him. I had actually thrown him to the floor and was beating him. I could not control myself; a part of me just wanted to continue beating as if all the anger and rage would then go away. I knew I had to stop, that if I didn't I was going to hurt him. I felt as if there was a war going on in my mind: part of me just wanted to keep hitting and beating him, and another part knew that somehow I had to stop, that I was going to hurt him severely if I didn't. I was terrified. Somehow, from somewhere, I got the strength and self-control to stop. I was so grateful that it wasn't too late. I held and rocked him and cried with him. I apologized to him and tried to reassure him of my love for him. I told him I had no right to do that to him and asked him to please forgive me. I tried to explain that I had lost my self-control. I didn't want him to take it personally, yet how could he help but take it that way. I think I knew what the next step had to be.

I got out the *Physicians' Desk Reference* and went into my room to read about Lithium. It confirmed the fact that Lithium was an element (actually I knew this from chemistry). It also stated that "Preclinical studies have shown that Lithium alters sodium transport in nerve and muscle cells and effects a shift toward intraneuronal metabolism of catecholamines, but the specific biochemical mechanism of Lithium action in mania is unknown." As I continued to read, I also found that "adverse reactions are seldom encountered" and that any "side effects are an inconvenience rather than a disabling condition." This information was basically the same as I had previously been told but it was reassuring: at least now I knew nothing was being kept from me.

When I gave my son his bath that night, I was horrified. He had marks all over his back and arms. I was so ashamed of myself for what I had done to him and I was even more scared than before, but for different reasons. The first was that because of my decision to question my doctor and discontinue using the Lithium, I had abused my son, whom I love very much. What effect would this have on him, and how would he feel about me now? Second, I was scared that when he went to school the next day, if anyone saw the marks and asked him what happened, he would tell them exactly how he got them. After all, I had taught him to tell the truth, that there was never a good excuse to lie. How I wished this one time that he would lie. How could I explain to anyone that I had abused my son for the sake of my ego. I knew then what my decision had to be, not only for me, but for the well being of my children.

My final decision, after I carefully reviewed my personal feelings and the actual facts, is that I will consistently take Lithium as long as my doctor feels I need it. I realize it will not prevent me from being myself, but rather will help me to show who I truly am. It does not prevent me from being upset when things go wrong, or stop the tears when I am sad (I am a very emotional person) but instead it helps me control my feelings.

I am now able to realize that I do have many friends and that people really do care about me. I am able to have better control of my emotions and feel much more secure about myself. My children and I get along much better and they do not have to worry about me having a tantrum. I only wish I had not questioned for so long, but had started the Lithium when it was first prescribed. My son might never have had to go through what he did, and I would not have to live with the memory of what I did to him. I realize that everything is not perfect, and that taking Lithium will not make things perfect, but our lives are on a much more even keel now.

Harriet was a most interesting and challenging patient. She was challenging in every sense of the word. She always had a series of questions that needed to be answered. She would find fault with whatever she was asked to do and would present a series of "logical" reasons why she couldn't or wouldn't do it. She had a number of college courses in psychology and that made her a self-styled psychologist. She used the little bit of knowledge she had extensively and consequently created an atmosphere that was at best testy.

She demanded that I ask questions, otherwise she would give inappropriate answers. If I were to ask her if she was still taking her medication, she would reply "yes." Actually, she was taking the medication haphazardly. The question she demanded to be asked was, "are you taking the medication every day and in exactly the manner prescribed?" Unless each and every question was posed to her in that fashion, she did not feel obliged to amplify or expand on her answers.

She recognized that she was playing games and that her behavior was self-destructive, however it continued for quite some time.

Harriet's history is not unusual. There are many patients who will not recognize their own need for Lithium Carbonate but will acknowledge the need in one of their offspring or parents. I have found this most often in professionals. I have many patients whose parents are professionals in the medical, dental, and legal fields. They are able to see and acknowledge the need of their family members to take Lithium Carbonate, but they are unwilling or unable to see their own need for it.

Harriet has a significant need for peer acceptance. Her peer group, especially those in school, suggested that she no longer take Lithium Carbonate. Since Lithium sensitive individuals are very pliable and need to please their peers, she acquiesced. Her clinical course deteriorated markedly. She once again had difficulty dealing with her children, and her anger and hostility began to mount.

This particular patient essay typifies the intensity to which people can lose control over their emotions. Anger, hostility, frustration, and the need for immediate gratification often give rise to explosive behavior patterns which can be overwhelmingly destructive to a family unit. Harriet gave clinical evidence and demonstrated the concerns we have about child abuse. Individuals who are Lithium deficient and who explode can do so to such an extent and degree that they cause significant harm to others. The battered child syndrome is one which we have seen in the households of many Lithium deficient individuals.

Individuals who are Lithium deficient often demand immediate responses to their wants and needs. Lithium sensitive individuals have a time orientation based on the present. They do not project well into the future, nor do they learn well from past history. Consequently, when a request is made of a child or a directive is given, it is expected that this request be adhered to immediately. When a person is Lithium sensitive, the chance is great that the child will also be Lithium sensitive. Therefore, a Lithium sensitive child may not readily acquiesce to the requests of another, or, if they do, while getting ready to carry it out, they may get easily diverted. As you can see, the situation becomes more and more complicated. There is often a significant clash of these personalities and the results often are either a physically or emotionally abused child.

Child abuse is an extremely common and serious situation.

It is of extreme importance that Lithium sensitive individuals be treated appropriately when a battered child is at stake. It is my understanding and opinion that the overwhelming majority of individuals who are considered battering parents, either emotionally or physically, are the way they are because of a Lithium deficiency syndrome. These individuals are easily recognized and should be put under treatment immediately. Harriet has not battered her child physically and only minimally, emotionally, since they are both taking Lithium Carbonate.

After months of her aforementioned behavior, Harriet once again resumed taking her Lithium Carbonate. It was only after her second clinical trial of Lithium Carbonate that she was finally willing to recognize that she did have a mineral deficiency and that much of her explosive behavior was out of her conscious control. It is crucial for patients to accept their illness and their course of treatment. Individuals must be willing to accept that they are Lithium deficient and must readily recognize the positive affects that Lithium Carbonate has on their personality. Without full recognition and acceptance, the treatment course is shaky at best. Consequently, continued patient education, explanation, and verification through literature and its presentation to individuals is a prime consideration in treatment. Without knowledgeable acceptance, one cannot expect individuals to take medication on a routine basis, even if it is a mineral.

Most patients who take Lithium Carbonate tend to readjust the dosage on their own. Most stop taking their medication sooner or later. Fortunately, most patients recognize the error of their ways and begin to take it again.

Chapter Eleven

THE END OF THE BEGINNING

For most of my life, I have been called a maverick. I have always had difficulty staying within the mainstream of my vocation. I think and believe that psychiatry is a viable and valuable medical specialty.

I do not feel or believe that the majority of individuals in the mental health professions are delivering the maximum quality of care to the maximum number of people. I do not believe that psychiatric treatment has to be long or costly.

When I present these viewpoints to individuals who are not in the mental health profession, fellow physicians, attorneys, roofers, welders, or waitresses, they agree. When I present these concepts to individuals who are practitioners in the mental health profession, there is usually intense disagreement. The conversation usually deteriorates to a tension filled attack on me and the concept. Fortunately, however, there are many individuals who, although they do not agree with me, are at least willing to listen and to discuss these issues, which allows both of us to reevaluate our positions.

While investigating people who are Lithium deficient and while developing the ideas and theories purported in this book, I have also been introduced to the whole concept of nutritional needs in psychiatric patients. Although I do not pretend to be well versed in the use of nutritional supplements such as vitamins and minerals (other than Lithium), I do recognize their usefulness in many patients. Lithium deficiency is not the only mineral or vitamin deficiency that gives rise to psychiatric illness, but it is, at least from my vantage point, the most common and most serious of the deficiencies. In the 1800s, Great Britain had a vast empire. As part of their thinking and philosophy, the British attempted to westernize Indonesia by polishing their rice. They felt that eating brown rice was rather barbaric and savage. Consequently, they polished the rice and presented this westernized, devitaminized rice to the populace. As time went on, the country, as a whole, became medically ill. The people developed a nutritional deficiency syndrome called beriberi. This was the end result of polishing rice. Unpolished rice contained adequate amounts of thiamine for the average individual's diet. Polishing the rice removed the thiamine. Unpolished rice was restored to the diet and the thiamine deficiency syndrome, beriberi, was eliminated. So, too, Lithium Carbonate has been eliminated from our water supply. Our very own population, as a result of the Industrial Revolution, has been the mainstay in the creation of this problem of Lithium deficiency. We are not advocating that people take extra Lithium. We are advocating that people replace the Lithium Carbonate which they used to get in their water supply with an external mineral supply. This supply can either be in tablet, capsule, or liquid form.

It is my fondest aspiration, and that of several others who work closely with me, to establish a Lithium clinic. It is our hope to be able to evaluate an individual in a matter of several hours and, when indicated, to begin a course of Lithium Carbonate therapy and education about the nature and quality of the problems, the use of Lithium, and how to deal with any and all complications. The clinic would be a place for individuals to come who think, after reading about their illness, that they do, in fact, have it. It is also a place where we hope to be able to train and educate other individuals in the mental health profession, as well as all physicians, in the application of minimal amounts of Lithium Carbonate. We hope this clinic will serve as an active and viable hub from which will sprout other centers with a similar orientation and theoretical philosophy. It is through the establishment of a clinic setting that the dissemination of knowledge and information will become more efficient and more widespread. It is through the establishment of a clinic that individuals who doubt the simplicity of psychiatric diagnosis and treatment can come, observe, and then formulate their own conclusions. Through the establishment of a clinic, data can be collected which will corroborate not only what has been said, but raise new questions and hopefully arrive at many new answers. We hope our clinic will be a meeting place for discussion, debate, and downright argument. It will be a place where both apostles and opponents are equally welcome.

Writing this book has given me the opportunity to meet many new people and to force myself to take a good, hard look at who I am, what I am, and have some idea or overview as to where I am going.

I recognize that a number of people may offhandedly discount these ideas and theories. Another group will discount these ideas and thoughts after reviewing them. However, I feel certain the overwhelming majority of individuals who are suffering from mood instability and inconsistency will recognize their personalities as described in this text.

Be sure when you seek proper medical and/or psychiatric evaluation that this evaluation is in fact performed by an individual or individuals who are versed in the Lithium deficiency syndromes and their multiple variations. This is truly an area of *caveat emptor*: may the buyer beware. Do not be concerned or afraid to thoroughly question any physician or mental health professional you choose to go to for your evaluation. Ask him what experience he has had in treating mood disorders, and specifically, how extensive is his knowledge and experience in the use of Lithium Carbonate, especially in low or subclinical dosages. Remember, the vast majority of individuals who require Lithium Carbonate do so in subclinical levels. It is not the blood test which guides the amount of Lithium that an individual needs, but rather their responsiveness to the medication. Be sure all of your questions are answered to your own satisfaction. Just as important, be sure the individual you deal with "feels right" for you and your needs.

After accomplishing all of the above, take your Lithium faithfully, sit back, and enjoy the subtle changes toward increased emotional stability.

You may also have recognized a significant number of individuals who suffer from mood instability, friends, co-workers, or perhaps even family members.

Caring for other individuals is often difficult. One of the more ticklish tasks is to help another individual seek proper medical (psychiatric) treatment for an emotional problem. Caring takes effort and not making the effort can be harmful, if not disastrous. I would like to implore each and every one of you who have recognized individuals who are Lithium deficient to present this information to them. Encourage them to make the effort to seek help so they can enjoy their life to its utmost. I recognize this task is one in which timing is very important and that the method you use to deliver this message to another individual is very important as well. An atmosphere of calm, quiet concern is a must, and the information should be presented in a low key but firm way. I also recognize that each and every individual to whom you present this information may not accept it in the light in which it is being offered, but at least you have offered it.

It is sometimes easier if you discuss another person—a mutual friend. When that individual's characteristics are enumerated and agreed upon, you may then be able to draw a parallel of that person and the person to whom you are speaking. The amount of resistance and the number of rationalizations used may astound you, but they may not. You may have used some of them yourself.

Change from within any system is slow and can be fraught with difficulty. Change must therefore be brought about both from within and from outside a given system. The mental health system is desperately in need of such change. You can help bring it about by insisting upon a return to basic mineral replacement. You can help recreate a relatively stable emotional environment in which society as a whole can flourish and grow.